Springer Theses

Recognizing Outstanding Ph.D. Research

For further volumes:
http://www.springer.com/series/8790

Aims and Scope

The series "Springer Theses" brings together a selection of the very best Ph.D. theses from around the world and across the physical sciences. Nominated and endorsed by two recognized specialists, each published volume has been selected for its scientific excellence and the high impact of its contents for the pertinent field of research. For greater accessibility to non-specialists, the published versions include an extended introduction, as well as a foreword by the student's supervisor explaining the special relevance of the work for the field. As a whole, the series will provide a valuable resource both for newcomers to the research fields described, and for other scientists seeking detailed background information on special questions. Finally, it provides an accredited documentation of the valuable contributions made by today's younger generation of scientists.

Theses are accepted into the series by invited nomination only and must fulfill all of the following criteria

- They must be written in good English.
- The topic should fall within the confines of Chemistry, Physics, Earth Sciences, Engineering and related interdisciplinary fields such as Materials, Nanoscience, Chemical Engineering, Complex Systems and Biophysics.
- The work reported in the thesis must represent a significant scientific advance.
- If the thesis includes previously published material, permission to reproduce this must be gained from the respective copyright holder.
- They must have been examined and passed during the 12 months prior to nomination.
- Each thesis should include a foreword by the supervisor outlining the significance of its content.
- The theses should have a clearly defined structure including an introduction accessible to scientists not expert in that particular field.

Nicola Salvi

Dynamic Studies Through Control of Relaxation in NMR Spectroscopy

Doctoral Thesis accepted by
Ecole Polytechnique Fédérale de Lausanne,
Switzerland

Author
Dr. Nicola Salvi
Department of Biological Chemistry
and Molecular Pharmacology
Harvard Medical School
Boston, MA
USA

Supervisor
Prof. Geoffrey Bodenhausen
Ecole Polytechnique Fédérale de Lausanne
Lausanne
Switzerland

ISSN 2190-5053 ISSN 2190-5061 (electronic)
ISBN 978-3-319-38335-4 ISBN 978-3-319-06170-2 (eBook)
DOI 10.1007/978-3-319-06170-2
Springer Cham Heidelberg New York Dordrecht London

Printed on acid-free paper

Springer is part of Springer Science+Business Media (www.springer.com)

Parts of this thesis have been published in the following journal articles:

1. Simone Ulzega, **Nicola Salvi**, Takuya F. Segawa, Fabien Ferrage, Geoffrey Bodenhausen, "Control of Cross Relaxation of Multiple-Quantum Coherences Induced by Fast Chemical Exchange under Heteronuclear Double-Resonance Irradiation". *ChemPhysChem* (**2011**), 12, 333.
2. **Nicola Salvi**, Simone Ulzega, Fabien Ferrage, Geoffrey Bodenhausen, "Time Scales of Slow Motions in Ubiquitin Explored by Heteronuclear Double Resonance". *Journal of the American Chemical Society* (**2012**), 134, 2481.

Haec mihi quae colitur violis pictura rosisque,
Quos referat voltus, Caediciane, rogas?
Talis erat Marcus mediis Antonius annis
Primus: in hoc iuvenem se videt ore senex.
Ars utinam more animumque effingere posset!
Pulchrior in terris nulla tabella foret.

—Martial, Epigrams, Book X, No. 32

Supervisor's Foreword

Proteins can be involved in a remarkably wide range of activities. It is generally assumed that the function of a protein is closely linked to its structure, a hypothesis known as the structure-function paradigm. In the last decades, structural biologists collected a great deal of evidence, mainly by using X-ray crystallography, to verify this paradigm. More recently it has been recognized that proteins must be flexible to perform their numerous tasks. Intrinsically Disordered Proteins (IDPs) offer an archetypal example: these proteins, many of which are of vital importance for living organisms, are unfolded in solution and, because they lack a well-defined structure under physiological conditions, they depend on their flexibility to become operational.

Nuclear Magnetic Resonance (NMR) spectroscopy is an ideal tool to study local motions of biomolecules. It can provide very detailed information at atomic resolution about frequencies and amplitudes of motions that span many orders of magnitude, from sub-nanosecond timescales to phenomena that occur over intervals of seconds or even hours. The thesis of Dr. Nicola Salvi offers a cogent view of some highly sophisticated NMR methods that he has developed in order to characterize the amplitudes and frequency ranges of local motions in biomolecules. His focus is on micro- to millisecond dynamics in proteins. These motions play an essential role since they can explain many of the remarkable properties of proteins and enable them to carry out all sorts of functions, from enzymatic catalysis to intermolecular recognition and signaling in cells.

Lausanne, March 2014 Geoffrey Bodenhausen

Abstract

In numerous biological processes that constitute the base of living organisms, protein function is fundamentally related to internal dynamics occurring on μs-ms timescales that can give rise to chemical exchange contributions to relaxation. In a heteronuclear two-spin system (e.g., ^{1}H-^{15}N), correlated motions of the two nuclei induce cross-relaxation between multiple-quantum coherences that can be quantified using new Heteronuclear Double Resonance (HDR) techniques.

An analytical model describes the effect of an applied radiofrequency field on the relaxation rate of interest in the presence of fast exchange, providing accurate information on the kinetics of correlated processes. Numerical simulations and experimental results confirm the validity of the model.

HDR and more conventional relaxation dispersion techniques are used to characterize the internal dynamics in several biological systems. Motions on similar timescales are detected in the two different binding surfaces in human ubiquitin and in the linker between them (Phe45), suggesting the presence of a possible global motion.

Additional applications of HDR give insights into the dynamics occurring in the KID-binding domain (KIX) of the CREB-binding protein. We identified the presence of exchange processes, faster than the one reported by Konrat and coworkers, outside the main helices of KIX. These fast motions are most likely the sign of conformational disorder within the native state, which may promote transition to the unfolded ensemble.

A combination of ^{15}N relaxation rates and magnetization transfer rates are used to provide a qualitative characterization of internal dynamics in Engrailed 2. While contribution of exchange processes in the ms timescale are small, fluctuations at sub-ms timescales are found to occur both in the unstructured part that contains important binding sites for other transcription factors, and at a few restricted locations in the structured homeodomain. Such motions are characterized by ^{15}N $R_{1\rho}$ relaxation dispersion. Our results reveal that the hexapeptide motif is characterized by complex dynamics that may be linked to its physiological function. Moreover, the timescale of motions in the homeodomain is reasonably close to that of the hexapeptide region. It is therefore tempting to suppose that transient contacts occur between the structured and unstructured regions.

Keywords BioNMR · Chemical exchange · Carr-Purcell-Meiboon-Gill (CPMG) · Rotating frame relaxation rates ($R_{1\rho}$) · Heteronuclear Double Resonance (HDR) · Ubiquitin · KIX · Engrailed

Acknowledgments

Over the past years I have received support and encouragement from a number of people without whom this thesis might not have been written, and to whom I am greatly indebted.

I thank Prof. Bodenhausen for giving me the opportunity to work in his lab and for his motivation, enthusiasm, and immense knowledge. His guidance helped in shaping me as a scientist and I am sure that I will benefit from this experience for the rest of my career.

Ringrazio il Dr. Simone Ulzega, meglio noto come "Ulega," non solo per avermi aiuto a penetrare nei meandri della teoria dell'HDR ma anche, forse soprattutto, per aver condiviso innumerevoli cene al Milan e in giro per il mondo, ore in palestra, pizze da Nando, i successi dell'Inter in Champions e le delusioni dell'Italia ai Mondiali. Grazie a lui ambientarsi a Losanna è stato molto più facile.

Besides my advisors, I would like to thank the rest of my thesis committee for taking the time to read my thesis and being present at my defense: Prof. Majed Chergui (EPFL), president of the jury, Prof. Mikael Akke (Lund University, Sweden), Prof. Lothar Helm (EPFL), Prof. Robert Konrat (University of Vienna, Austria).

I am indebted to Dr. Fabien Ferrage, who made available his support in a number of ways, from Skype-enabled basic hands-on training in my first days as a Ph.D. student to proofreading of this manuscript. I used to say that I owe him half of what I know about NMR and the percentage is constantly increasing.

When I started to work on HDR, I had the pleasure to share my frustrations and my doubts in front of the 800 with Dr. Takuya F. Segawa. I am truly indebted to him not only for his collaboration to scientific projects but also for being our helpful librarian and for taking care of many everyday tasks at the LRMB.

Nel corso di questi anni ho ereditato dalla Dr. ssa Mariachiara Verde il progetto di ricerca su cui questa tesi si fonda, la scrivania, il compagno di ufficio e metà della mia giuria. L'impatto positivo che ha avuto sul mio lavoro è evidente e le sono profondamente riconoscente.

Besides HDR, I spent some time working on long-lived states together with Aurélien Bornet and Roberto Buratto. I learnt a lot and I had lots of fun thanks to them and I would like to show my gratitude to them for making possible one of the best works I had the privilege to co-author. Special thanks to Aurélien for having had the patience to take me to Walt Disney World. Roberto rimarrà il mio fotografo ufficiale durante le conferenze.

During my Ph.D. studies I had the pleasure to share the office with two postdocs. At my arrival I was welcomed by Dr. Bikash Baishya. I had the opportunity to appreciate his kindness and his spontaneity. I am grateful to him for making our office a wonderful working space. Il Dr. Diego Carnevale è subentrato a Bikash nel ruolo di tenere alto il morale nell'ufficio. Lo ringrazio per questo, per tante serate al Great Escape, e per avermi introdotto, con alterne fortune, alla musica, ai Soliti Idioti e al solid-state NMR. È stato molto interessante per me scoprire un modo completamente nuovo di lavorare con gli spin e sarà un privilegio per me poter essere co-autori di una pubblicazione.

Aurélien, Bikash, Diego, Simone and Tak were not the only people I appreciated at the LRMB: Dr. Puneet Ahuja, Dr. Marc Caporini, Quentin Chappuis, Dr. Srinivas Chithalapalli, Dr. Sami Jannin, Daniele Mammoli, Jonas Milani, Dr. A. J. Perez-Linde, Dr. Pascal Miéville, Dr. Riddhiman Sarkar, Prof. Paul Vasos, Dr. Veronika Vitzthum, Basil Vuichoud and Shutao Wang contributed, in different ways, to my work and they are an important part of my life in Lausanne.

Synergies with the group in Paris made my work at the EPFL even more enjoyable and fruitful. Besides Fabien, I am indebted to Dr. Philippe Pelupessy for many insightful discussions. It was a pleasure to collaborate with Dr. Rafal Augustyniak on the characterization of dynamics in Engrailed. I had the privilege to meet Cyril Charlier and Shahid Khan and to have with them serious discussions, more funny ones, and a handful of beers. I would like to thank them for being good friends and excellent collaborators.

I am grateful to Prof. Robert Konrat, Prof. Martin Tollinger (University of Innsbruck, Austria), and Dr. Sven Brüschweiler (University of Innsbruck, Austria) for providing us with KIX sample and giving me the opportunity to work on such a challenging and interesting biological system.

I would like to show my gratitude to Martial Rey and to Pascal for maintaining our machines and for being my primary resource to solve hardware issues. I was able to run experiments because there are qualified and competent people that make the spectrometers run.

It is a pleasure to thank Béatrice Bliesener-Tong and Anne Lene Odegaard for the precious work they carry out in secret in such an efficient way that it is not even possible to quantify its impact on my research.

Ringrazio la mia famiglia per il costante supporto e affetto e per la vicinanza al di là di ogni barriera geografica. Sapere una persona cara lontana e in una situazione stressante può essere difficile, perciò il loro contributo a questo lavoro è inestimabile.

Ringrazio i miei amici italiani di ogni età per il supporto, l'affetto, l'incoraggiamento e la stima chemi hanno fatto sempre sentire speciale: Alessandro, Andrea, Angela, Demian, Eleonora, Giampiero, Greta, P. Lino, Maria Grazia, Paolo, don Ramon, Simona, Simonetta e tanti altri che ometto per mancanza di spazio, ma che sanno che porto sempre nel cuore.

Infine, un ringraziamento speciale va a Fanny. Non ho avuto la fortuna di farmi accompagnare da lei durante tutti i quattro anni passati a Losanna, ma posso dire che i mesi passati in sua compagnia sono stati di gran lunga i più felici, intensi e ricchi di soddisfazioni di tutto il dottorato.

Lausanne, May 2013 Nicola Salvi

Contents

Chapter 1
Introduction

1.1 Proteins: Function, Structure and Dynamics

Proteins are essential components of all living organisms, responsible for a remarkably wide range of biological functions, from the catalysis of biochemical reactions to the regulation of transport phenomena, from physical movements to gene expression.

An important feature of proteins is their specificity of function: an enzyme will bind a specific substrate to catalyze a specific chemical reaction, a given antibody will a bind specific antigen, etc. Besides, many of these processes can be regulated by the binding of small ligands or other biomolecules, sometimes in a site of the protein distant from the one responsible for its function.

Given this functional variety, one would expect a corresponding variety in the structure of biomolecules: it is not by chance that *structural* biology has had a huge impact on our understanding of biological processes. Indeed, X-ray crystallography, nuclear magnetic resonance (NMR) spectroscopy and electron microscopy have been very powerful in determining the three-dimensional structures of proteins and protein complexes. Since Perutz solved the first X-ray structure of Haemoglobin in 1960 [1], X-ray crystallography has elucidated more than 70,000 protein structures. Currently, protein complexes up to several Mega daltons (10^6 au $= 1$ MDa) can be investigated by X-ray crystallography [2]. More recently, cryo electron microscopy has been used to obtain low-resolution (about 10 Å) images of very large systems [3, 4].

Since the seminal works by Wüthrich and co-workers [5, 6], NMR spectroscopy has been developed in parallel to X-ray crystallography as an alternative approach to investigate the structure of proteins. The advent of three-dimensional NMR spectroscopy [7, 8] and the use of isotopic labeling techniques [9–11] allowed the study of biomolecules of higher weight and complexity. Nowadays, using advanced labeling strategies, it is possible to study systems up to hundreds of kDa using NMR spectroscopy [12]. X-ray crystallography still performs better than NMR in terms

N. Salvi, *Dynamic Studies Through Control of Relaxation in NMR Spectroscopy*, Springer Theses, DOI: 10.1007/978-3-319-06170-2_1,

of the size of protein structures that can be solved. However, in contrast to X-ray crystallography, NMR spectroscopy is able to provide information about dynamics that are crucial to the understanding of protein function.

Indeed, quoting Gregorio Weber [13],

> The protein molecule model resulting from the X-ray crystallographic observations is a "platonic" protein, well removed in its perfection from the kicking and screaming "stochastic" molecule that we infer must exist in solution.

In other words, proteins are inherently dynamic systems, and their motions cover an enormous range of time scales from 10^{-12} to 10^{2} s and more. Since motions are an intrinsic property of proteins, evolution used it to improve their functions. It is therefore fundamental to elucidate the dynamics-function relationship. For a detailed atomic understanding of a protein's function, the 3D structure *and* the accurate description of its dynamic behavior are necessary.

Besides NMR spectroscopy, protein dynamics can be studied by a number of different techniques, such as optically time-resolved methods after temperature jumps [14, 15], infrared spectroscopy [16, 17] and fluorescence correlation spectroscopy [18, 19]. In addition, promising techniques such as time-resolved X-ray crystallography [20, 21], two-dimensional infrared spectroscopy [22], temperature-jump fluorescence-based methods [23] and temperature-jump Fourier-transformation infrared spectroscopy (FT-IR) [24] are emerging to access protein dynamics, in some cases with atomic resolution. However, each one of the mentioned techniques suffers from various specific bottlenecks. We are not going to discuss these limitations here, preferring to underline that NMR spectroscopy has several unique and important advantages over other experimental techniques:

1. it covers a **wide time range**, from femtoseconds to hours;
2. only **minimal interference** with the native protein is required: except for replacing the more abundant isotopes ^{12}C and ^{14}N with the NMR-observable ^{13}C and ^{15}N, chemical modifications of the protein are not necessary and the experiments can be carried out in a nearly physiological environment, or even directly in a cell;
3. in contrast to time-resolved techniques, kinetics can be measured at **equilibrium**.

1.2 NMR Methods to Study Protein Dynamics

In addition to the advantages outlined in the previous sections, NMR spectroscopy provides many different ways to investigate protein dynamics, as summarized in Fig. 1.1. Over the last decades the NMR community has witnessed a relentless development of methods to exploit the information provided by these probes. These experimental techniques have been extensively reviewed and the reader is referred to [25–29] for more details. Here we aim at summarizing the general principles. A deeper discussion of the methods used in this work will be presented in Chap. 3. Following the scheme of Fig. 1.1, we can identify at least four different classes of techniques:

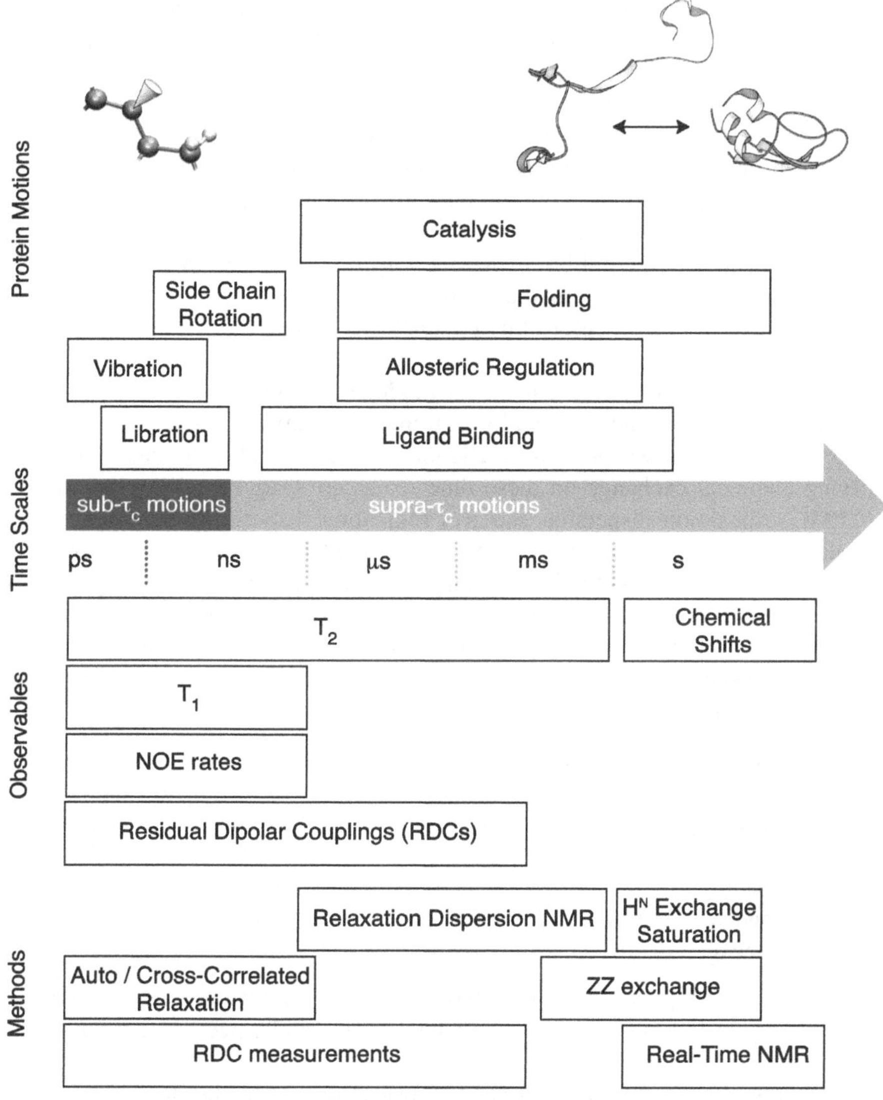

Fig. 1.1 Time scales of protein motions, NMR observables and methods to probe them

- conformational motions on time scales faster than or comparable to the rotational correlation time τ_c (which for biomolecules is usually several ns or longer) stochastically modulate the dipole–dipole (DD), the chemical shift anisotropy (CSA) and quadrupolar Hamiltionans. These stochastic modulations induce nuclear spin relaxation, and therefore longitudinal (T_1) and transverse (T_2) relaxation rates and Nuclear Overhauser Effect (NOE) rates provide information about ps-ns motions

[26, 29]. Usually backbone amide NH groups are the focus of studies to investigate protein backbone motions, [11] and methyl groups to probe side chain dynamics [30–32]. Relaxation of carbonyl ^{13}C spins provides an additional probe of the dynamics of the backbone peptide plane [33]. Relaxation rates are linear combinations of the spectral density function $J(\omega)$ at some characteristic frequencies of the spin system. Therefore, discrete values of $J(\omega)$ can be obtained from relaxation data. This procedure is called "spectral density mapping" [34, 35]. The model-free formalism of Lipari and Szabo allows one to extract a generalized order parameter that is a measure of the equilibrium distribution of orientations in a molecular reference frame of a vector defining the orientation of the unique axis of the relaxation interaction [36, 37]. Therefore this parameter is a source of information about sub-τ_c motions;

- motions on μs-ms time scales can induce modulations of isotropic chemical shifts, resulting in an additional contribution to transverse relaxation. Therefore, T_2's are also sensitive to slower processes. The main experimental techniques for quantifying chemical exchange on these time scales are Carr-Purcell-Meiboom-Gill (CPMG) relaxation dispersion and $R_{1\rho}$ relaxation dispersion, most commonly applied to ^{1}H, ^{13}C, ^{15}N, and ^{31}P spins in biological macromolecules [28, 29]. Relaxation dispersion methods offer unique advantages. Indeed, by using analytical models that describe the effect of the repetition rate of π pulses (in the case of CPMG experiments) or of the amplitude of the spin-lock field (in the case of $R_{1\rho}$ relaxation dispersion) it is possible to extract a wealth of information with atomic resolution about:

  1. the **kinetics** of the chemical exchange or conformational change: it is possible to determine directly and accurately the time scale of the phenomenon;
  2. the **thermodynamics**: relaxation dispersion data allow one to measure the populations of two (or more) conformations in exchange;
  3. the **structure** of the otherwise *invisible* weakly populated excited state in exchange with the main conformer, by measuring the chemical shift differences between the two states.

  Most of the work presented in this thesis is based on relaxation dispersion techniques. We shall discuss the theory and the experimental aspects of these methods in Chaps. 3 and 4, respectively;

- Dipolar couplings between spins have an orientational angular dependence with respect to the external magnetic field. They are generally averaged out in solution-state NMR spectroscopy because of the tumbling of the protein and the isotropic distribution of its orientations. However, proteins can be aligned in solution by using slightly anisotropic media. Therefore an anisotropic distribution of orientations is introduced, and so is a small fraction (about 0.1 %) of the total dipolar coupling interaction, i.e. the residual dipolar coupling (RDC) [38, 39]. RDC's are fixed to the molecular frame independently of the tumbling of the molecule. Because of this property, they are sensitive to motions that are either faster or slower than the rotational tumbling correlation time of the molecule. Two similar approaches, known as the RDC-based model-free approach [40] and direct interpretation of dipolar couplings (DIDC) [41] have been used. Both rely on RDC data acquired in at least five different alignment media. The main outcome is a set of RDC-based order parameters, that, in contrast to the Lipari-Szabo order parameters, reflect both sub-τ_c *and* supra-τ_c motions. Therefore, a comparison between the two reveals the presence of exchange processes slower than τ_c;

- in the slow-exchange limit, i.e., when the exchange rate k_{ex} is smaller than the chemical shift difference $\Delta\omega$ between the conformers, a distinct resonance peak for each conformer appears in the NMR spectrum. The dynamic process will influence both the intensities and the line widths of these resonances. Dynamic processes can be studied, for instance, by starting from non-equilibrium conditions and monitoring the change of the intensities in time. Specially in the case of binding processes, the chemical shift of the species can be monitored while changing the concentration of one of the binding partners in titration experiments. A fitting of the titration curve can be used to extract the dissociation constants that are related to the association and dissociation rates of the complex.

1.3 Outline of the Present Thesis

As shown in Fig. 1.1, many interesting biological processes, such as ligand binding, allosteric regulation, protein folding and enzyme catalysis of chemical reactions, occur on μs-ms time scales, i.e. on time scales that are best studied by relaxation dispersion NMR. Therefore, the focus of my PhD studies has been the development of relaxation dispersion techniques and their application to the investigation of biological systems.

The **theoretical aspects** of relaxation dispersion methods for studying internal dynamics in proteins are discussed in Chaps. 2 and 3. The Bloch equations that describe the behavior of a nuclear spin in a magnetic field and under radio-frequency (*rf*) irradiation are introduced before discussing the basics of Redfield's relaxation theory. A brief introduction to Average Hamiltonian Theory (AHT) and Average Liouvillian Theory (ALT) will present some theoretical tools that are helpful to describe the effect of complex pulse sequences on the coherent spin evolution and on the nuclear spin relaxation. In Chap. 3, these theoretical concepts are used to describe relaxation dispersion methods and to develop analytical models that are important for the analysis of experimental data. Special attention is given to the Heteronuclear Double Resonance (HDR) method that we have developed.

Chapter 4 deals with **experimental aspects** of relaxation-based methods. The pulse sequences used in the present work are discussed, as well as several important considerations about the analysis of the data. Similarly to Chap. 3, the HDR method is the most original topic of this part of the thesis.

The **experimental results** obtained during my PhD studies are reported in Chap. 5. Relaxation dispersion data, complemented with insights provided by other relaxation-based methods, are used to characterize the internal dynamics in

(a) *ubiquitin*, a small (8.5 kDa) regulatory protein that is found in almost all tissues of eukaryotic organisms and has been studied for several decades. Nevertheless, many aspects of its dynamics are still unclear;
(b) the KID-binding (*KIX*) domain of the CREB-binding protein (CBP), that is involved in various signaling pathways and in the binding to a number of partners, that are often intrinsically unstructured in isolation but undergo folding upon binding to KIX, thus underlining the prominent role of coupled folding and binding in the fine-tuning of protein-protein interactions [42];
(c) *Engrailed 2*, a transcription factor of the class of homeoproteins that contains both structured and disordered regions.

References

1. Perutz MF, Rossmann MG, Cullis AF, Muirhead H, Will G, North ACT (1960) Nature 185:416–422
2. Ban N, Nissen P, Hansen J, Moore PB, Steitz TA (2000) Science 289:905–920
3. Halic M, Blau M, Becker T, Mielke T, Pool MR, Wild K, Sinning I, Beckmann R (2006) Nature 444:507–511
4. Halic M, Gartmann M, Schlenker O, Mielke T, Pool MR, Sinning I, Beckmann R (2006) Science 312:745–747
5. Wagner G, Wüthrich K (1978) Nature 275:247–248
6. K Wüthrich, NMR of proteins and nucleic acids. Wiley Interscience, 1986
7. Griesinger C, Sørensen OW, Ernst RR (1987) J Am Chem Soc 109:7227–7228
8. Oschkinat H, Griesinger C, Kraulis PJ, Ernst RR, Gronenborg AM, Clore GM (1988) Nature 332:374–376
9. Oh BH, Westler WM, Darba P, Markley JL (1988) Science 240:908–911
10. Marion D, Kay LE, Sparks SW, Torchia DA, Bax A (1989) J Am Chem Soc 111:1515–1517
11. Kay LE, Torchia DA, Bax A (1989) Biochemistry 29:4659–4667
12. Baldwin AJ, Walsh P, Hansen DF, Hilton GR, Benesch JL, Sharpe S, Kay LE (2012) J Am Chem Soc 134:15343–15350
13. Weber G (1975) Adv Protein Chem 29:1–83
14. Ervin J, Sabelko J, Gruebele M (2000) J Photochem Photobiol B 54:1–15
15. Fierz B, Joder K, Krieger F, Kiefhaber T (2007) Methods Mol Biol 350:169–187
16. Hamm P, Zurek M, Mantele W, Meyer M, Scheer H, Zinth W (1995) Proc Natl Acad Sci USA 92:1826–1830
17. Snow CD, Nguyen H, Pande VS, Gruebele M (2002) Nature 420:102–106
18. Oehlenschlager F, Schwille P, Eigen M (1996) Proc Natl Acad Sci USA 93:12811–12816
19. Kettling U, Koltermann A, Schwille P, Eigen M (1998) Proc Natl Acad Sci USA 95:1416–1420
20. Srajer V, Teng T, Ursby T, Pradervand C, Ren Z, Adachi S, Schildkamp W, Bourgeois D, Wulff M, Moffat K (1996) Science 274:1726–1729
21. Ihee H, Lorenc M, Kim TK, Kong QY, Cammarata M, Lee JH, Bratos S, Wulff M (2005) Science 309:1223–1227
22. Bredenbeck J, Helbing J, Nienhaus K, Nienhaus GU, Hamm P (2007) Proc Natl Acad Sci USA 104:14243–14248
23. Ma H, Gruebele M (2005) Proc Natl Acad Sci USA 102:2283–2287
24. Chung HS, Ganim Z, Jones KC, Tokmakoff A (2007) Proc Natl Acad Sci USA 104:14237–14242
25. Palmer AG (1997) Curr Opin Struct Biol 7:732

26. Kay LE (1998) Nat Struct Biol 5:513
27. Fischer MWF, Majumdar A, Zuiderweg ERP (1998) Prog Nucl Magn Reson Spectrosc 33:207
28. Palmer AG, Kroenke CD, Loria JP (2001) Methods Enzymol 204:204
29. Palmer AG (2004) Chem Rev 104:3623
30. Millet O, Muhandiram DR, Skrynnikov NR, Kay LE (2002) J Am Chem Soc 124:6439
31. Skrynnikov NR, Millet O, Kay LE (2002) J Am Chem Soc 124:6439
32. Mittermaier A, Kay LE (2006) Science 312:224
33. Wang T, Cai S, Zuiderweg ERP (2003) J Am Chem Soc 125:8639
34. Peng JW, Wagner G (1992) J Magn Reson 98:308
35. Peng JW, Wagner G (1992) Biochemistry 31:8571
36. Lipari G, Szabo A (1982) J Am Chem Soc 104:4546
37. Lipari G, Szabo A (1982) J Am Chem Soc 104:4559
38. Tolman JR, Flanagan JM, Kennedy MA, Prestegard JH (1995) Proc Natl Acad Sci USA 92:9279–9283
39. Tjandra N, Bax A (1997) Science 278:1111
40. Meiler J, Prompers JJ, Peti W, Griesinger C, Brüschweiler R (2001) J Am Chem Soc 123:6098–6107
41. Tolman JR (2002) J Am Chem Soc 124:12020–12030
42. Tollinger M, Kloiber K, Agoston B, Dorigoni C, Lichtenecker R, Schmid W, Konrat R (2006) Biochemistry 45:8885–8893

Chapter 2
Theoretical Principles

In this chapter we shall present a phenomenological description of nuclear spin dynamics, including relaxation, in a magnetic field (i.e., the Bloch equations) as well as a more accurate description using the principles of quantum mechanics (Redfield relaxation theory). We shall then use these concepts to develop analytical models to describe relaxation dispersion experiments and to derive analytical functions that allow one to extract information from the experimental data.

2.1 The Bloch Equations

In this section we shall follow the approach of [1]. The semiclassical vector model developed by Felix Bloch in [2] describes the behavior of an ensemble of noninteracting spin-1/2 nuclei in a static magnetic field but fails in describing more complex systems, where a quantum mechanical approach is required. However, because the basic concepts and terminology are still used in modern NMR, we shall briefly present the Bloch equations.

2.1.1 Equations Describing Spin Dynamics in the Absence of Relaxation

The evolution of the bulk magnetization vector $\mathbf{M}(t)$ in the presence of a magnetic field $\mathbf{B}(t)$ is described by

$$\frac{d\mathbf{M}(t)}{dt} = \mathbf{M}(t) \times \gamma \mathbf{B}(t) \tag{2.1}$$

where γ is the gyromagnetic ratio of the nuclei of interest. Equation 2.1 is valid in the laboratory frame. It is convenient to rewrite it in a frame that rotates with angular velocity ω around the z-axis, defined by the unit vector $\mathbf{k}$:

N. Salvi, *Dynamic Studies Through Control of Relaxation in NMR Spectroscopy*, Springer Theses, DOI: 10.1007/978-3-319-06170-2_2,

$$\left[\frac{d\mathbf{M}(t)}{dt}\right]_{rot} = \left[\frac{d\mathbf{M}(t)}{dt}\right]_{lab} + \mathbf{M}(t) \times \omega\mathbf{k}$$
$$= \mathbf{M}(t) \times \left[\gamma\mathbf{B}(t) + \omega\mathbf{k}\right]. \tag{2.2}$$

The two equations can be actually written in the same form, provided that $\mathbf{B}(t)$ is replaced by an *effective* field defined as

$$\mathbf{B}_{eff} = \mathbf{B}(t) + \frac{\omega}{\gamma}\mathbf{k}. \tag{2.3}$$

For a static field $|\mathbf{B}(t)| = B_0$, the effective field is zero if ω is chosen to correspond to the *Larmor frequency*

$$\omega_0 = -\gamma B_0, \tag{2.4}$$

and the bulk magnetization appears to be stationary. In other terms, in the absence of other fields the bulk magnetization simply precesses at the Larmor frequency around the magnetic field that defines the z-direction of the laboratory and rotating frames.

The magnetic component of a radio-frequency (*rf*) field that is linearly polarized along the x-axis of the laboratory frame is

$$\begin{aligned}\mathbf{B}_{rf}(t) &= 2B_1 \cos\left(\omega_{rf}t + \phi\right)\mathbf{i} \\ &= B_1\{\cos\left(\omega_{rf}t + \phi\right)\mathbf{i} + \sin\left(\omega_{rf}t + \phi\right)\mathbf{j}\} \\ &\quad + B_1\{\cos\left(\omega_{rf}t + \phi\right)\mathbf{i} - \sin\left(\omega_{rf}t + \phi\right)\mathbf{j}\},\end{aligned} \tag{2.5}$$

where B_1 is the amplitude of the applied *rf* field, ω_{rf} its angular (or carrier) frequency, ϕ its phase, and $\mathbf{i}$ and $\mathbf{j}$ are unit vectors defining the x- and y-axes, respectively. In the second equality of the previous equation, we decomposed the linearly polarized *rf* field into two circularly polarized components, with opposite directions of rotation around the z-axis. Only the component rotating in the same sense as the magnetic moment can interact with the nuclear spins. The counter-rotating field produces only a small effect, known as Bloch-Siegert shift, that is proportional to $(B_0/2B_1)^2$, and can be neglected for our purposes. Therefore, one obtains

$$\mathbf{B}_{rf}(t) = B_1\{\cos\left(\omega_{rf} + \phi\right)\mathbf{i} + \sin\left(\omega_{rf} + \phi\right)\mathbf{j}\}. \tag{2.6}$$

It is useful then to rewrite Eq. 2.1 in a rotating frame where the *rf* field is time-independent, i.e. a frame that rotates at angular frequency ω_{rf}:

$$\frac{d\mathbf{M}^{rff}(t)}{dt} = \mathbf{M}^{rff}(t) \times \gamma\mathbf{B}^{rff}, \tag{2.7}$$

where the effective field in the *rf* frame is

$$\mathbf{B}^{rff} = B_1\{\cos\phi\mathbf{i}^{rff} + \sin\phi\mathbf{j}^{rff}\} + \Delta B_0\mathbf{k}^{rff}, \tag{2.8}$$

and

$$\Delta B_0 = -\frac{\Omega}{\gamma}, \tag{2.9}$$

where we used the definition of the offset $\Omega = \omega_0 - \omega_{rf}$. The magnitude of the effective field is

$$B^{rff} = \sqrt{B_1^2 + \Delta B_0^2} = \frac{B_1}{\sin\theta}, \tag{2.10}$$

where θ indicates the angle through which the frame is tilted with respect to z-axis of the laboratory frame:

$$\tan\theta = \frac{B_1}{\Delta B_0} = \frac{\omega_1}{\Omega}. \tag{2.11}$$

In general, we will use this tilted frame in the following discussion, unless otherwise stated. For the sake of clarity, we will use a simplified notation, dropping the superscript *rff*.

2.1.2 Empirical Description of NMR Relaxation

Using the equations of the previous section, the magnetization would evolve around the z-axis freely and forever. This is of course not the case, because the experimental practice shows that the thermal equilibrium is restored after some time. In other words, nuclear spins are subject to relaxation. Bloch [2] proposed to introduce two processes to account for such relaxation. The first mechanism accounts for the return of the population difference across the Zeeman transition to Boltzmann equilibrium, i.e. for the z-component of the magnetization to go back to equilibrium. This process, known as longitudinal or spin-lattice relaxation, can be described by a first-order rate expression:

$$\frac{dM_z(t)}{dt} = R_1\left[M_0 - M_z(t)\right], \tag{2.12}$$

in which $R_1 = 1/T_1$ is the longitudinal relaxation rate constant and M_0 is the magnitude of the equilibrium magnetization, which lies entirely on the z-axis. According to Eq. 2.12, the longitudinal magnetization returns to equilibrium in an exponential fashion.

Another first-order rate process, known as transverse or spin-spin relaxation, was introduced to model the decay of the transverse magnetization in the transverse plane:

$$\begin{aligned} \frac{dM_x(t)}{dt} &= -R_2 M_x(t), \\ \frac{dM_y(t)}{dt} &= -R_2 M_y(t), \end{aligned} \tag{2.13}$$

where $R_2 = 1/T_2$ is the transverse relaxation rate constant. This second process accounts for the mutual dephasing of the spins in the xy-plane.

The Bloch equations are obtained combining Eqs. 2.1, 2.12 and 2.13:

$$\begin{aligned}\frac{dM_x(t)}{dt} &= \gamma\,[\mathbf{M}(t)\times\mathbf{B}(t)]_x - R_2 M_x(t)\,,\\ \frac{dM_y(t)}{dt} &= \gamma\,[\mathbf{M}(t)\times\mathbf{B}(t)]_y - R_2 M_y(t)\,,\\ \frac{dM_z(t)}{dt} &= \gamma\,[\mathbf{M}(t)\times\mathbf{B}(t)]_z - R_1\left[M_z(t) - M_0\right].\end{aligned} \tag{2.14}$$

These equations describe the evolution of the magnetization in a static magnetic field and in the absence of an applied *rf* field. A more general expression can be obtained in a rotating reference frame including the effect of an *rf* pulse (see Eqs. 2.6 and 2.7). In a convenient matrix form, one obtains

$$\frac{\mathrm{d}}{\mathrm{d}t}\begin{bmatrix}M_x(t)\\M_y(t)\\M_z(t)\end{bmatrix} = \begin{bmatrix}-R_2 & -\Omega & \omega_1\sin\phi\\ \Omega & -R_2 & -\omega_1\cos\phi\\ -\omega_1\sin\phi & \omega_1\cos\phi & -R_1\end{bmatrix}\begin{bmatrix}M_x(t)\\M_y(t)\\M_z(t)\end{bmatrix} + R_1 M_0\begin{bmatrix}0\\0\\1\end{bmatrix}. \tag{2.15}$$

In many cases of practical interest the equation can be simplified even further. For instance, during free precession, that is in the absence of *rf* pulses, $\omega_1 = 0$ and 2.15 simplifies to

$$\frac{\mathrm{d}}{\mathrm{d}t}\begin{bmatrix}M_x(t)\\M_y(t)\\M_z(t)\end{bmatrix} = \begin{bmatrix}-R_2 & -\Omega & 0\\ \Omega & -R_2 & 0\\ 0 & 0 & -R_1\end{bmatrix}\begin{bmatrix}M_x(t)\\M_y(t)\\M_z(t)\end{bmatrix} + R_1 M_0\begin{bmatrix}0\\0\\1\end{bmatrix}. \tag{2.16}$$

If the pulse is short enough, i.e. its duration $\tau_p << T_1, T_2$, it is possible, in a good approximation, to neglect relaxation contributions to the trajectory of the magnetization:

$$\frac{\mathrm{d}}{\mathrm{d}t}\begin{bmatrix}M_x(t)\\M_y(t)\\M_z(t)\end{bmatrix} = \begin{bmatrix}0 & -\Omega & \omega_1\sin\phi\\ \Omega & 0 & -\omega_1\cos\phi\\ -\omega_1\sin\phi & \omega_1\cos\phi & 0\end{bmatrix}\begin{bmatrix}M_x(t)\\M_y(t)\\M_z(t)\end{bmatrix}. \tag{2.17}$$

As mentioned earlier, the Bloch equations fail at describing systems of interacting spins. Several extensions to Bloch model to account for such interactions, such as the Solomon Eq. [3], have been proposed. However, to be useful for practical applications, it is necessary to use the semiclassical approach of Bloch, Wangsness and Redfield [4, 5].

2.2 Chemical Exchange Effects

In the absence of scalar coupling interactions, chemical exchange processes are described by an extension of the Bloch equations (see Sect. 2.1). In the case of a first-order chemical reaction or, equivalently, a two-site chemical exchange process, the kinetic rate laws are written in matrix form as

$$\frac{\mathrm{d}}{\mathrm{d}t}\begin{bmatrix}[A_1](t)\\ [A_2](t)\end{bmatrix}=\begin{bmatrix}-k_1 & k_{-1}\\ k_1 & -k_{-1}\end{bmatrix}\begin{bmatrix}[A_1](t)\\ [A_2](t)\end{bmatrix}, \tag{2.18}$$

in which k_1 (k_{-1}) is the rate constant for the forward (reverse) reaction. In general, in the case of a set of N coupled reactions, one has

$$\frac{\mathrm{d}\mathbf{A}(t)}{\mathrm{d}t}=\mathbf{KA}(t), \tag{2.19}$$

in which the elements of the rate matrix $\mathbf{K}$ are given by

$$K_{ij}=k_{ji}, \tag{2.20}$$

$$K_{ii}=-\sum_{\substack{j=1\\ j\neq i}}^{N}k_{ij}. \tag{2.21}$$

Modified Bloch equations, known as McConnell equations, can be derived for such a system:

$$\begin{aligned}
\frac{\mathrm{d}\mathbf{M}_{jx}(t)}{\mathrm{d}t}&=\gamma\left(1-\sigma_j\right)\left[\mathbf{M}_j(t)\times\mathbf{B}(t)\right]_x-R_{2j}M_{jx}(t)+\sum_{k=1}^{N}K_{jk}\mathbf{M}_{kx}(t),\\
\frac{\mathrm{d}\mathbf{M}_{jy}(t)}{\mathrm{d}t}&=\gamma\left(1-\sigma_j\right)\left[\mathbf{M}_j(t)\times\mathbf{B}(t)\right]_y-R_{2j}M_{jy}(t)+\sum_{k=1}^{N}K_{jk}\mathbf{M}_{ky}(t),\\
\frac{\mathrm{d}\mathbf{M}_{jz}(t)}{\mathrm{d}t}&=\gamma\left(1-\sigma_j\right)\left[\mathbf{M}_\mathbf{j}(t)\times\mathbf{B}(t)\right]_z-R_{1j}\left[M_{jz}(t)-M_{j0}\right]+\sum_{k=1}^{N}K_{jk}\mathbf{M}_{kz}(t).
\end{aligned} \tag{2.22}$$

The above equations can be generalized to higher-order reactions by introducing the pseudo-first order rate constants:

$$k_{ij}=\frac{\zeta_{ij}(t)}{[A_i](t)}, \tag{2.23}$$

where $\zeta_{ij}(t)$ is the rate constant for the conversion of the ith species into the jth one.

The McConnell equations can be solved analogously to the Bloch equations. Three different regimes emerge from the set of solutions to these equations:

1. in the case of *slow exchange*, i.e. when the exchange rate is smaller than the chemical shift difference between the two sites, two resonance lines are observed;
2. as the exchange rate increases, the lines broaden and, when the rate is of the order of the chemical shift difference, the lines coalesce (*intermediate exchange* or *coalescence*);
3. if the rate is further increased, the system is in *fast exchange* and a single narrow resonance line is observed at the average of the chemical shifts.

2.3 Bloch-Wangsness-Redfield Theory

In this model, also known as Redfield theory, a quantum mechanical description of the system is derived, while describing the surroundings (i.e., the heat bath or lattice) in a classical way. The main limitation of this approximation is that the energy levels are predicted to be equally populated at equilibrium. Therefore, the theory is formally valid only in the high-temperature limit, which is a very good approximation at room temperature. At finite temperatures, corrections are required to ensure that the correct equilibrium populations are reached. However, these corrections are significant only in the case of very low temperatures. In our discussion, we shall follow the account of [1].

2.3.1 The Master Equation

Let us write the Hamiltonian as a sum of terms that act only on the spin system ($\mathcal{H}_0$) and a stochastic part, $\mathcal{H}_1(t)$, that couples the spin system to the lattice:

$$\mathcal{H}(t) = \mathcal{H}_0 + \mathcal{H}_1(t)\,. \tag{2.24}$$

In the above expression the absence of applied *rf* fields was implicitly assumed, thus $\mathcal{H}_0$ is time-independent. In other words, a time-dependent perturbation $\mathcal{H}_1(t)$ is superimposed onto the main time-independent Hamiltonian $\mathcal{H}_0$.

The corresponding Liouville equation of motion, describing the evolution of the density operator $\sigma(t)$, is

$$\frac{\mathrm{d}\sigma(t)}{\mathrm{d}t} = -i\left[\mathcal{H}_0 + \mathcal{H}_1(t), \sigma(t)\right] = -i\left\{\hat{L}_0 + \hat{L}_1(t)\right\}\sigma(t)\,, \tag{2.25}$$

in which $\hat{L}(t) = [\mathcal{H}(t)]$, is the commutation superoperator or Liouvillian.

It is convenient to remove the explicit dependence on $\mathscr{H}_0$ by rewriting the density operator in a new reference frame, called the *interaction frame*:

$$\sigma^T(t) = \exp\left(i\hat{L}_0 t\right)\sigma(t) = \exp(i\mathscr{H}_0 t)\,\sigma(t)\exp(-i\mathscr{H}_0 t). \tag{2.26}$$

Transforming also the stochastic Hamiltonian,

$$\mathscr{H}_1^T(t) = \hat{L}_0\mathscr{H}_1(t) = \exp(i\mathscr{H}_0 t)\,\mathscr{H}_1(t)\exp(-i\mathscr{H}_0 t), \tag{2.27}$$

it is possible to rewrite Eq. 2.25 in the interaction frame:

$$\frac{\mathrm{d}\sigma^T(t)}{\mathrm{d}t} = -i\left[\mathscr{H}_1^T(t), \sigma^T(t)\right] = -i\hat{L}_1^T(t)\,\sigma^T(t). \tag{2.28}$$

In mathematical terms, the transformation into the interaction frame is isomorphous to the rotating-frame transformation. However, there are some marked differences between the two. Indeed, in the rotating frame the *rf* Hamiltonian is time-independent and the interactions contained in $\mathscr{H}_0$ are retained; on the other hand, in the interaction frame, $\mathscr{H}_0$ is not (explicitly) active, whereas the time dependence of $\mathscr{H}_1^T(t)$ is retained. This means that it is possible to apply both the transformation sequentially.

Several assumptions are required to solve Eq. 2.28:

1. The ensemble average of $\mathscr{H}_1^T(t)$ is zero. Any time-dependent fluctuations that do not vanish upon averaging are to be included in the time-independent Hamiltonian;
2. $\sigma^T(t)$ and $\mathscr{H}_1^T(t)$ are not correlated, thus it is possible to take the ensemble average of the fluctuations of the Hamiltonian and of the quantum states independently;
3. $\tau_c \ll t \ll 1/R$, where τ_c is the correlation time relevant for $\mathscr{H}_1^T(t)$ and R is the relevant relaxation rate constant;
4. in order for the system to relax towards thermal equilibrium, $\sigma^T(t)$ has to be replaced by $\sigma^T(t) - \sigma_0$, in which σ_0 is the density operator at equilibrium. By definition, one has $\sigma_0^T = \sigma_0$.

Using this assumption, the right-hand term in Eq. 2.28 can be replaced by an integral:

$$\frac{\mathrm{d}\sigma^T(t)}{\mathrm{d}t} = -\int_0^\infty \mathrm{d}\tau\,\overline{[\mathscr{H}_1^T(t), [\mathscr{H}_1^T(t-\tau)}, \sigma^T(t) - \sigma_0]], \tag{2.29}$$

where the overbar represents the ensemble average. In this equation the third assumption allows the integral to run to infinity; because we assumed that the fluctuations of the Hamiltonian are not correlated with the density matrix, we could calculate the ensemble average over the stochastic Hamiltonians independently from $\sigma^T(t)$.

In order to be able to transform Eq. 2.29 back to the laboratory frame, the stochastic Hamiltonian has to be decomposed as the sum of random functions of spatial variables $F_k^q(t)$ and tensor spin operators $\mathbf{A}_k^q$:

$$\mathscr{H}_1(t) = \sum_{q=-k}^{k} (-1)^q F_k^{-q}(t)\, \mathbf{A}_k^q. \tag{2.30}$$

The tensor spin operators are chosen to be spherical tensor operators because of their transformations properties under rotations. For the Hamiltonians of interest in NMR spectroscopy, the rank of the tensor k is one or two. These operators can be further decomposed as a sum of basis operators:

$$\mathbf{A}_k^q = \sum_p \mathbf{A}_{kp}^q, \tag{2.31}$$

where the components $\mathbf{A}_{kp}^q$ have the following property:

$$\hat{L}_0 \left\{ \mathbf{A}_{kp}^q \right\} = \left[\mathscr{H}_0, \mathbf{A}_{kp}^q \right] = \omega_p^q \mathbf{A}_{kp}^q. \tag{2.32}$$

In other words, $\mathbf{A}_{kp}^q$ are eigenfunctions of the Hamiltonian commutation superoperator with eigenfrequencies ω_p^q. The index p here is used to distinguish between spin operators with the same order q but different eigenfrequencies. Furthermore, Eq. 2.32 implies the following property:

$$\exp\left(i\hat{L}_0 t\right) \mathbf{A}_{kp}^q = \exp\left(i\mathscr{H}_0 t\right) \mathbf{A}_{kp}^q \exp\left(-i\mathscr{H}_0 t\right) = \exp\left(i\omega_p^q t\right) \mathbf{A}_{kp}^q, \tag{2.33}$$

which defines also the transformation of $\mathbf{A}_{kp}^q$ in the interaction frame:

$$\mathbf{A}_k^{qT} = \exp\left(i\mathscr{H}_0 t\right) \mathbf{A}_k^q \exp\left(-i\mathscr{H}_0 t\right) = \sum_p \mathbf{A}_{kp}^q \exp\left(i\omega_p^q t\right). \tag{2.34}$$

Substituting Eqs. 2.30 and 2.34 in Eq. 2.29, one obtains

$$\frac{\mathrm{d}\sigma^T(t)}{\mathrm{d}t} = -\sum_{q,q'} \sum_{p,p'} (-1)^{q+q'} \exp\left\{ i\left(\omega_p^q + \omega_{p'}^{q'}\right) t \right\} \left[\mathbf{A}_{kp'}^{q'}, \left[\mathbf{A}_{kp}^q, \sigma^T(t) - \sigma_0 \right] \right]$$
$$\times \int_0^\infty \overline{F_k^{-q'}(t)\, F_k^{-q}(t-\tau) \exp\left(i\omega_p^q \tau\right) \mathrm{d}\tau}. \tag{2.35}$$

If $q' \neq -q$, the two random processes $F_k^{-q'}(t)$ and $F_k^{-q}(t)$ are assumed to be statistically independent, which causes the ensemble average to vanish, unless $q' = -q$. Thus,

$$\frac{d\sigma^T(t)}{dt} = -\sum_{q=-k}^{k}\sum_{p,p'}\exp\left\{i\left(\omega_p^q - \omega_{p'}^q\right)t\right\}\left[\mathbf{A}_{kp'}^{-q},\left[\mathbf{A}_{kp}^{q},\sigma^T(t)-\sigma_0\right]\right]$$
$$\times\int_0^\infty \overline{F_k^q(t)\,F_k^{-q}(t-\tau)\exp\left(i\omega_p^q\tau\right)}d\tau. \quad (2.36)$$

The equation above can be further simplified by noticing that terms in which $|\,\omega_p^q - \omega_{p'}^q\,| \gg 0$ oscillate much faster than the typical time scales of relaxation phenomena, and therefore do not affect the evolution of the density matrix. Furthermore, in the absence of degenerate eigenfrequencies, terms in Eq. 2.36 are not vanishing only if $p = p'$. Therefore,

$$\frac{d\sigma^T(t)}{dt} = -\sum_{q=-k}^{k}\sum_{p}\left[\mathbf{A}_{kp}^{-q},\left[\mathbf{A}_{kp}^{q},\sigma^T(t)-\sigma_0\right]\right]\int_0^\infty \overline{F_k^q(t)\,F_k^{-q}(t-\tau)\exp\left(i\omega_p^q\tau\right)}d\tau. \quad (2.37)$$

The terms $\overline{F_k^q(t)\,F_k^{-q}(t-\tau)}$ are known as *correlation functions*. The real part of the integral in Eq. 2.37 is the *power spectral density function* $j^q(\omega)$:

$$\begin{aligned} j^q(\omega) &= 2\,\mathrm{Re}\left\{\int_0^\infty \overline{F_k^q(t)\,F_k^{-q}(t-\tau)}\exp(i\omega\tau)\,d\tau\right\} \\ &= \mathrm{Re}\left\{\int_{-\infty}^\infty \overline{F_k^q(t)\,F_k^{-q}(t-\tau)}\exp(i\omega\tau)\,d\tau\right\} \\ &= \mathrm{Re}\left\{\int_{-\infty}^\infty \overline{F_k^q(t)\,F_k^{-q}(t+\tau)}\exp(i\omega\tau)\,d\tau\right\}. \end{aligned} \quad (2.38)$$

The equation above shows that the power spectral density is an even function of τ. Furthermore, it is an even function of ω as well. On the other hand, the imaginary part of the integral in Eq. 2.37,

$$\begin{aligned} k^q(\omega) &= \mathrm{Im}\left\{\int_0^\infty \overline{F_k^q(t)\,F_k^{-q}(t-\tau)}\exp(-i\omega\tau)\,d\tau\right\} \\ &= \mathrm{Im}\left\{\int_0^\infty \overline{F_k^q(t)\,F_k^{-q}(t+\tau)}\exp(-i\omega\tau)\,d\tau\right\}, \end{aligned} \quad (2.39)$$

is an odd function of ω.

In the high-temperature limit, the equilibrium density matrix is proportional to $\mathscr{H}_0$. Thus, using Eq. 2.33, the double commutator $\left[\left[\mathbf{A}_{kp}^{-q}, \mathbf{A}_{kp}^{q}\right], \sigma_0\right] = 0$. Using this property in Eq. 2.37, one obtains

$$\frac{\mathrm{d}\sigma^T(t)}{\mathrm{d}t} = -\frac{1}{2}\sum_{q=-k}^{k}\sum_{p}\left[\mathbf{A}_{kp}^{-q}, \left[\mathbf{A}_{kp}^{q}, \sigma^T(t) - \sigma_0\right]\right] j^q\left(\omega_p^q\right) + i\sum_{q=0}^{k}\sum_{p}\left[\left[\mathbf{A}_{kp}^{-q}, \mathbf{A}_{kp}^{q}\right], \sigma^T(t)\right] k^q\left(\omega_p^q\right). \tag{2.40}$$

By transforming the above equation back to the laboratory frame, the *Liouville-von Neumann* differential equation for the density equation is obtained:

$$\frac{\mathrm{d}\sigma(t)}{\mathrm{d}t} = -i\left[\mathscr{H}_0, \sigma(t)\right] - i\left[\Delta, \sigma(t)\right] - \hat{\Gamma}\left(\sigma(t) - \sigma_0\right), \tag{2.41}$$

in which the *relaxation superoperator* is

$$\hat{\Gamma} = \frac{1}{2}\sum_{q=-k}^{k}\sum_{p}\left[\mathbf{A}_{kp}^{-q}, \left[\mathbf{A}_{kp}^{q}, \;\right]\right] j^q\left(\omega_p^q\right). \tag{2.42}$$

Δ is the dynamic frequency shift operator that accounts for second-order frequency shifts of the resonance lines, known as dynamic frequency shifts:

$$\Delta = -\sum_{q=0}^{k}\sum_{p} k^q\left(\omega_p^q\right)\left[\mathbf{A}_{kp}^{-q}, \mathbf{A}_{kp}^{q}\right]. \tag{2.43}$$

This term can be incorporated into the Hamiltonian to obtain the final result, known as *master equation*:

$$\frac{\mathrm{d}\sigma(t)}{\mathrm{d}t} = -i\left[\mathscr{H}_0, \sigma(t)\right] - \hat{\Gamma}\left(\sigma(t) - \sigma_0\right). \tag{2.44}$$

In the calculation of relaxation rates it is often convenient to expand Eq. 2.44 in terms of the basis operators used to expand the density operator:

$$\frac{\mathrm{d}b_r(t)}{\mathrm{d}t} = \sum_{s}\left\{-i\Omega_{rs} b_s(t) - \Gamma_{rs}\left[b_s(t) - b_{s0}\right]\right\}, \tag{2.45}$$

where Ω_{rs} are characteristic frequencies defined as

$$\Omega_{rs} = \frac{\langle \mathbf{B}_r \mid [\mathscr{H}_0, \mathbf{B}_s]\rangle}{\langle \mathbf{B}_r \mid \mathbf{B}_s\rangle}, \tag{2.46}$$

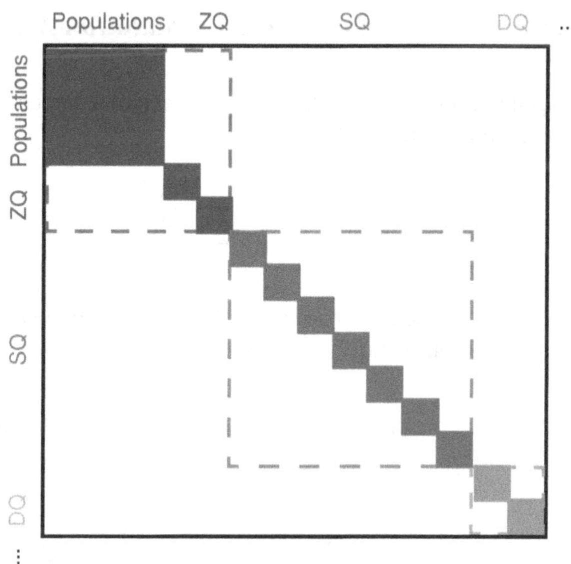

Fig. 2.1 Redfield kite. In the absence of degenerate transitions, only the elements contained in the *solid blocks* can be non-zero, i.e. cross-relaxation is possible between populations, while the coherences relax independently. In the case of degenerate transitions, additional non-zero elements can be found inside the *dashed blocks* between coherences of the same order. *ZQ* zero quantum, *SQ* single quantum, *DQ* double quantum

Γ_{rs} are the rate constants for relaxation between the operators $\mathbf{B}_s$ and $\mathbf{B}_r$ (and vice versa, since $\Gamma_{rs} = \Gamma_{sr}$ for normalized basis operators)

$$\begin{aligned}\Gamma_{rs} &= \frac{\left\langle \mathbf{B}_r \mid \hat{\Gamma}\mathbf{B}_s \right\rangle}{\langle \mathbf{B}_r \mid \mathbf{B}_r \rangle} \\ &= \frac{1}{2}\sum_{q=-k}^{k}\sum_{p}\left\{\frac{\left\langle \mathbf{B}_r \mid \left[\mathbf{A}_{kp}^{-q}, \left[\mathbf{A}_{kp}^{q}, \mathbf{B}_s\right]\right]\right\rangle}{\langle \mathbf{B}_r \mid \mathbf{B}_r \rangle}\right\} j^q\left(\omega_p^q\right),\end{aligned} \tag{2.47}$$

and finally $b_r(t)$ results from the application of the relevant projection superoperator onto the density matrix:

$$b_r(t) = \frac{\langle \mathbf{B}_r \mid \sigma(t) \rangle}{\langle \mathbf{B}_r \mid \mathbf{B}_r \rangle}. \tag{2.48}$$

The diagonal elements Γ_{rr} are *auto*-relaxation rates, while off-diagonal elements Γ_{rs} are *cross*-relaxation rates. Because we assumed that only terms satisfying $q = -q'$ give non-zero contributions to Eq. 2.35, cross-relaxation can occur only between operators with the same coherence order. In addition, because of the secular approximation in Eq. 2.37, cross-relaxation between off-diagonal terms is forbidden in the absence of degenerate transitions. These two features give rise to a characteristic block shape in the relaxation superoperator, known as Redfield kite (see Fig. 2.1).

2.3.2 Interference Between Relaxation Mechanisms

In general there is more than one process that can cause relaxation. In this case, one can generalize Eq. 2.30 to

$$\mathscr{H}_1(t) = \sum_m \sum_{q=-k}^{k} (-1)^q F_{mk}^{-q}(t)\, \mathbf{A}_{mk}^q, \tag{2.49}$$

in which the index m runs over all active mechanisms. Using the above equation, a generalization of Eq. 2.47 can be written:

$$\begin{aligned}\Gamma_{rs} &= \frac{1}{2} \sum_m \sum_{q=-k}^{k} \sum_p \left\{ \frac{\left\langle \mathbf{B}_r \mid \left[\mathbf{A}_{mkp}^{-q}, \left[\mathbf{A}_{mkp}^q, \mathbf{B}_s \right] \right] \right\rangle}{\langle \mathbf{B}_r \mid \mathbf{B}_r \rangle} \right\} j^q\left(\omega_p^q\right) \\ &+ \frac{1}{2} \sum_{\substack{m,n \\ m \neq n}} \sum_{q=-k}^{k} \sum_p \left\{ \frac{\left\langle \mathbf{B}_r \mid \left[\mathbf{A}_{mkp}^{-q}, \left[\mathbf{A}_{nkp}^q, \mathbf{B}_s \right] \right] \right\rangle}{\langle \mathbf{B}_r \mid \mathbf{B}_r \rangle} \right\} j_{mn}^q\left(\omega_p^q\right) \\ &= \sum_m \Gamma_{rs}^m + \sum_{\substack{m,n \\ m \neq n}} \Gamma_{rs}^{mn},\end{aligned} \tag{2.50}$$

in which the cross-correlated spectral density was used, defined as

$$j_{mn}^q = \mathrm{Re}\left\{ \int_{-\infty}^{\infty} \overline{F_{mk}^q(t)\, F_{nk}^{-q}(t+\tau)} \exp(-i\omega\tau)\, \mathrm{d}\tau \right\}. \tag{2.51}$$

In other words, according to Eq. 2.50, in a spin system where more than one stochastic Hamiltonian is present, we can have relaxation due to only one mechanism, with relaxation rate constant Γ_{rs}^m, and relaxation arising from the interference (or cross-correlation) between different mechanisms, with relaxation rate constant Γ_{rs}^{mn}. The latter is possible only if there is some degree of correlation between the mth and nth mechanisms, i.e., if $F_{mk}^q(t)$ and $F_{nk}^q(t)$ are correlated.

2.3.3 Relaxation in the Rotating Frame

In the presence of an applied *rf* field, a transformation into a rotating frame, where the time dependence of the *rf* Hamiltonian $\mathscr{H}_{rf}(t)$ is removed, has to precede the transformation into the interaction frame. When the Zeeman interaction is the dominant term in $\mathscr{H}_0$, the interaction frame is a doubly rotating tilted frame. Thus, the frequencies used as arguments of the spectral density function should be replaced by $\omega_p^q + \omega_p^{q(rf)}$, where the latter is defined by

$$\sum_{i=1}^{K} \omega_{rf,i} \left[I_{zi}, \mathbf{A}_{kp}^{q} \right] = \omega_{p}^{q(rf)} \mathbf{A}_{kp}^{-q}, \tag{2.52}$$

in which $\omega_{rf,i}$ is the frequency of the rotating frame for the ith spin and K is the number of irradiated spins.

However, in most of the cases $\omega_1 \tau_c \ll 1$, where ω_1 is the applied *rf* amplitude and τ_c is the rotational correlation time of the molecule of interest. Therefore $j^q \left(\omega_p^q + \omega_p^{q(rf)} \right) \approx j^q \left(\omega_p^q \right)$. Using this approximation, it is possible to compute approximate relaxation rate constants by transforming the operators in the tilted frame to the laboratory frame before using Eq. 2.47:

$$\Gamma'_{rs} = \frac{\left\langle \mathbf{U}^{-1} \mathbf{B}'_r \mathbf{U} \mid \hat{\Gamma} \left\{ \mathbf{U}^{-1} \mathbf{B}'_s \mathbf{U} \right\} \right\rangle}{\left\langle \mathbf{B}'_r \mid \mathbf{B}'_r \right\rangle}. \tag{2.53}$$

For instance, if an *rf* field is applied with x-phase, the transformation is defined as a rotation around the y-axis:

$$\mathbf{U} = \exp \left\{ i \sum_{i=1}^{K} \theta_i I_{yi} \right\}, \tag{2.54}$$

in which $\theta_i = \omega_1 / \Omega_i$ is the tilt angle.

The autorelaxation rate $R_1(\theta_i)$ or $R_{1\rho}$ is given by

$$R_{1\rho} = R_1 \cos^2 \theta_i + R_2 \sin^2 \theta_i. \tag{2.55}$$

Operators that do not commute with the Hamiltonian in the rotating frame decay rapidly as a consequence of *rf* inhomogeneities: therefore, if a continuous-wave field is applied, only operators which do not evolve in the rotating frame have to be considered, i.e. longitudinal operators and zero-quantum coherences. If the rf field is phase- or amplitude-modulated in order to suppress the effect of offsets and *rf* inhomogeneities, more quantum states have to be considered and the effective average rate constant has to be obtained by averaging the instantaneous rate constant over the trajectory followed by the operator under the influence of the rotating-frame Hamiltonian (see Sects. 2.4 and 2.5).

2.3.4 Spectral Density Functions

As shown by Hubbard [6], in the high-temperature limit, the description of the relaxation properties of the system requires only one spectral density function, because the following identity holds:

Table 2.1 Spatial functions for relaxation mechanisms

Interaction	$c(t)$
Dipolar	$-\sqrt{6}\,(\mu_0/4\pi)\,\hbar\gamma_I\gamma_S r_{IS}(t)^{-3}$
CSA	$\Delta\sigma\gamma_I B_0/\sqrt{3}$
Quadrupolar	$e^2qQ/[4\hbar I(2I-1)]$

An axially symmetric chemical shift tensor is assumed, with $\sigma_{zz} = \sigma_{\|}$, $\sigma_{xx} = \sigma_{yy} = \sigma_{\perp}$ and $\Delta\sigma = \sigma_{\|} - \sigma_{\perp}$. The electric field gradient tensor is assumed to be axially symmetric, with principal values $V_{zz} = eq$ and $V_{xx} = V_{yy}$. Q is the nuclear quadrupole moment and e is the electron charge

$$j^q(\omega) = (-1)^q j^0(\omega) \equiv (-1)^q j(\omega). \tag{2.56}$$

Tensor operators of rank 2 can be used to describe the relaxation mechanisms of interest. The random functions $F_2^0(t)$ in Eq. 2.39 can be factored as

$$F_2^0(t) = c_0(0)\, Y_2^0[\Omega(t)]. \tag{2.57}$$

Therefore, the spectral density function can be written as

$$\begin{aligned} j(\omega) &= \mathrm{Re}\left\{\int_{-\infty}^{\infty} \overline{c_0(t)\, c_0(t+\tau)\, Y_2^0[\Omega(t)]\, Y_2^0[\Omega(t+\tau)]}\, \exp(-i\omega\tau)\, \mathrm{d}\tau\right\} \\ &= \mathrm{Re}\left\{\int_{-\infty}^{\infty} C(\tau)\exp(-i\omega\tau)\, \mathrm{d}\tau\right\}, \end{aligned} \tag{2.58}$$

with the stochastic correlation function:

$$C(\tau) = \overline{c_0(t)\, c_0(t+\tau)\, Y_2^0[\Omega(t)]\, Y_2^0[\Omega(t+\tau)]}. \tag{2.59}$$

In the above equations, $c_0(t)$ is a function of physical constants and spatial variables (see Table 2.1), $Y_2^0[\Omega(t)]$ is a modified second-order spherical harmonic function (see Table 2.2) of the polar angles in the laboratory frame $\Omega(t)$. The polar angles are used to define the orientation in the laboratory frame of a vector that points in the principal direction for the interaction. The most important feature of spectral density functions is that, as the molecules tumble in solution, the oscillating magnetic fields that cause relaxation are not distributed in a homogeneous way over all frequencies. The power spectral density function can be used to measure the probability of motions with frequency between ω and $\omega + \mathrm{d}\omega$.

For a rigid spherical molecule, the spatial function is time-independent (i.e., $c(t) = c_0$), and therefore $J(\omega) = d_{00}J(\omega)$, in which $d_{00} = c_0^2$. The orientational spectral density function used here is defined as

Table 2.2 Modified second-order spherical harmonics

q	Y_2^q
0	$\left(3\cos^2\theta - 1\right)/2$
1	$-\sqrt{3/2}\sin\theta\cos\theta e^{i\phi}$
2	$\sqrt{3/8}\sin^2\theta e^{i2\phi}$

$$J(\omega) = \mathrm{Re}\left\{\int_{-\infty}^{\infty} C_{00}^2(\tau)\exp(-i\omega\tau)\,\mathrm{d}\tau\right\}, \tag{2.60}$$

in which the orientational correlation function

$$C_{00}^2(\tau) = \overline{Y_2^0[\Omega(t)]\,Y_2^0[\Omega(t+\tau)]} \tag{2.61}$$

was used. For instance, in the case of isotropic rotational diffusion of a rigid rotor, the orientational correlation function is

$$C_{00}^2(\tau) = \frac{1}{5}e^{-\tau/\tau_c}, \tag{2.62}$$

in which τ_c is the correlation time, which depends on the size of the molecule, the viscosity of the solution, and the temperature. The corresponding spectral density function is

$$J(\omega) = \frac{2}{5}\frac{\tau_c}{\left(1+\omega^2\tau_c^2\right)}. \tag{2.63}$$

The functional form in Eq. 2.63 is a Lorentzian. Therefore, because the value of a Lorentzian is almost constant for $\omega^2\tau_c^2 < 1$, if the molecular motion is rapid enough to satisfy $\omega_p^{(q)2}\tau_c^2 \ll 1$ (i.e., τ_c is short enough), $J\left(\omega_p^q\right) \approx J(0)$ is a good approximation. We shall refer to this limit as *extreme narrowing* regime. On the contrary, if the motion is very slow, i.e. $\omega_p^{(q)2}\tau_c^2 \gg 1$, then $J\left(\omega_p^q\right) \propto \omega_p^{(q)-2}\tau_c$, a limit known as *spin diffusion* (or slow tumbling) regime.

Both overall rotational Brownian motions and relative motions of nuclei in a molecular reference frame contribute to the modulation of local magnetic fields. In the case of isotropic rotational diffusion, to a very good approximation, the total correlation function can be factored as

$$C(\tau) = C_O(\tau)\,C_I(\tau), \tag{2.64}$$

in which the correlation function for overall motions $C_O(\tau)$ is given by Eqs. 2.61 and 2.62, whereas the correlation function for internal motions has to be computed directly from Eq. 2.59 assuming a model of intramolecular motions.

Table 2.3 Tensor operators for the dipolar interaction

q	p	A^q_{2p}	ω_p
0	0	$\left(2/\sqrt{6}\right) I_z S_z$	0
0	1	$-1/\left(2\sqrt{6}\right) I^+ S^-$	$\omega_I - \omega_S$
1	0	$-(1/2) I_z S^+$	ω_S
1	1	$-(1/2) I^+ S_z$	ω_I
2	0	$(1/2) I^+ S^+$	$\omega_I + \omega_S$

2.3.5 *Relaxation Mechanisms*

In the following, we will limit ourselves to intramolecular dipolar, anisotropic chemical shift, quadrupolar and scalar coupling interactions. Intramolecular paramagnetic relaxation can be described by the same Hamiltonian as dipolar interactions, the only difference being that the interaction is between the nucleus and an unpaired electron. All other relaxation mechanisms are of no practical interest in the case of biomolecules. In particular, for spin-1/2 nuclei in diamagnetic biological molecules, the dipolar and the anisotropic chemical shift mechanisms are by far the dominant mechanisms.

2.3.5.1 Dipolar Relaxation

In the case of intramolecular dipolar relaxation for a an IS spin system, the terms $\mathbf{A}^q_{2p}$ are given in Table 2.3. The relaxation rate constants can be calculated using Eq. 2.47 and are given in Table 2.4. It is worth pointing out that R_1 has a maximum for $\omega_0 \tau_c = 1$, whereas R_2 increases monotonically with τ_c.

If the two spins are weakly coupled, the longitudinal relaxation is unaffected by the scalar interaction because both I_z and S_z commute with the scalar coupling Hamiltonian. The expressions in Table 2.4 are therefore still valid.

As far as the transverse relaxation is concerned, the in-phase I^+ (S^+) term evolves into the anti-phase operator $2I^+S_z$ ($2I_zS_+$) under the effect of the scalar interaction. If the evolution is faster than the relaxation processes, an average relaxation constant is measured because the magnetization is rapidly exchanging between in-phase and anti-phase terms.

2.3.5.2 Chemical Shift Anisotropy and Quadrupolar Relaxation

Local fields that lie at the origin of chemical shifts are in general anisotropic. Therefore, the chemical shift is best described as a tensor. The reorientation of this tensor with respect to the laboratory frame induces a time-varying magnetic field on the nucleus and, consequently, relaxation. The chemical-shift anisotropy (CSA) relaxation is significant for ^{13}C, ^{15}N and ^{31}P, while it is negligible for protons, and it has a quadratic dependence on the strength of the static magnetic field.

Table 2.4 Relaxation rate constants for *IS* dipolar interaction

Coherence level	Operator	Relaxation rate constant
Populations	I_z	$(d_{00}/4)\{J(\omega_I-\omega_S)+3J(\omega_I)+6J(\omega_I+\omega_S)\}$
	S_z	$(d_{00}/4)\{J(\omega_I-\omega_S)+3J(\omega_I)+6J(\omega_I+\omega_S)\}$
	$I_z \leftrightarrow S_z$	$(d_{00}/4)\{-J(\omega_I-\omega_S)+6J(\omega_I+\omega_S)\}$
0	$2I_zS_z$	$(3d_{00}/4)\{J(\omega_I)+J(\omega_S)\}$
	ZQ_x, ZQ_y	$(d_{00}/8)\{2J(\omega_I-\omega_S)+3J(\omega_I)+3J(\omega_S)\}$
±1	I^+, I^-	$(d_{00}/8)\{4J(0)+J(\omega_I-\omega_S)+3J(\omega_I)+6J(\omega_S)+6J(\omega_I+\omega_S)\}$
	S^+, S^-	$(d_{00}/8)\{4J(0)+J(\omega_I-\omega_S)+3J(\omega_S)+6J(\omega_I)+6J(\omega_I+\omega_S)\}$
	$2I^+S_z, 2I^-S_z$	$(d_{00}/8)\{4J(0)+J(\omega_I-\omega_S)+3J(\omega_I)+6J(\omega_I+\omega_S)\}$
	$2I_zS^+, 2I_zS^-$	$(d_{00}/8)\{4J(0)+J(\omega_I-\omega_S)+3J(\omega_S)+6J(\omega_I+\omega_S)\}$
±2	DQ_x, DQ_y	$(d_{00}/8)\{3J(\omega_I)+3J(\omega_S)+12J(\omega_I+\omega_S)\}$

$d_{00}=(\mu_0/4\pi)^2\hbar^2\gamma_I^2\gamma_S^2r_{IS}^{-6}$

Table 2.5 Tensor operators for the CSA interaction

q	p	A_{2p}^q	ω_p
0	0	$\left(2/\sqrt{6}\right)I_z$	0
1	0	$-(1/2)I^+$	ω_I

Table 2.6 Tensor operators for the quadrupolar interaction

q	p	A_{2p}^q	ω_p
0	0	$\left(1/2\sqrt{6}\right)\left[4I_z^2-I^+I^--I^-I^+\right]$	0
1	0	$-(1/2)\left[I_zI^++I^+I_z\right]$	ω_I
2	0	$(1/2)I^+I^+$	$2\omega_I$

Nuclei with $I>1/2$ also have a nuclear electric quadrupole moment, which is a measure of how much the nuclear charge distribution departs from a spherical one. A relaxation pathway is provided by the interaction of the quadrupole moment with local oscillating electric field gradients, generated by the electrons.

The terms $\mathbf{A}_{2p}^q$ for the CSA and quadrupolar interactions are given in Tables 2.5 and 2.6, respectively. Longitudinal and transverse relaxation rate constants are given in Table 2.7, where axially symmetrical CSA and quadrupolar tensors were assumed for a spin-1 nucleus.

2.3.5.3 Scalar Relaxation

In the presence of a scalar interaction, the local magnetic field experienced by spin S (e.g., a nitrogen-15) depends on the value of the J-coupling constant with spin I (e.g., a proton). The magnetic field is time-dependent in the following two scenarios:

1. the value of the J-coupling constant is time-dependent (*scalar relaxation of the first kind*). This can happen in the case of transitions of the spin system between environments where the coupling constant assumes different values;

Table 2.7 CSA and quadrupolar relaxation rate constants

Rate constant	CSA	Quadrupolar
R_1	$d_{00}J(\omega_I)$	$3d_{00}\{J(\omega_I)+2J(2\omega_I)\}$
R_2	$(d_{00}/6)\{4J(0)+3J(\omega_I)\}$	$(3d_{00}/2)\{3J(0)+5J(\omega_I)+2J(2\omega_I)\}$

CSA $d_{00}=\left(\sigma_\parallel-\sigma_\perp\right)^2\omega_I^2/3$

Quadrupolar $d_{00}=\left[e^2qQ/(4\hbar)\right]^2$

2. the state of spin I varies very rapidly (*scalar relaxation of the second kind*). This is the case if the nucleus relaxes rapidly or if it is involved in fast chemical exchange.

Expressions for contributions to relaxation rate constants from scalar relaxation are given by [7]:

$$R_1^{sc}=\frac{2A^2}{3}S(S+1)\frac{\tau_2}{1+(\omega_I-\omega_S)^2\tau_2^2}; \tag{2.65}$$

$$R_2^{sc}=\frac{A^2}{3}S(S+1)\left[\frac{\tau_2}{1+(\omega_I-\omega_S)^2\tau_2^2}+\tau_1\right]. \tag{2.66}$$

For scalar relaxation of the first kind, $A=2\pi(p_1p_2)^{1/2}(J_1-J_2)$, in which J_1 and J_2 are the scalar coupling constants, p_1 and p_2 the populations of the two sites, and $\tau_1=\tau_2=\tau_e$ is the exchange time constant. In the case of relaxation of the second kind, $A=2\pi J_{IS}$ and τ_1 and τ_2 are the spin-lattice and spin-spin relaxation time constants for spin S.

2.4 Average Hamiltonian Theory

In the following section, we will adopt the approach of [8].

The principle on which Average Hamiltonian Theory (AHT) rests is that, in the case of a periodic time-dependent Hamiltonian, it is possible, under suitable conditions, to describe the evolution of the spin system with good accuracy considering the average effect of the Hamiltonian over a cycle of its oscillations [9]. For a time-independent system, the evolution operator can be written in an exponential form. Therefore, in the time-dependent case, for the Schrödinger equation

$$\frac{d\sigma(t)}{dt}=-i\mathcal{H}(t)\sigma(t), \tag{2.67}$$

a solution expressed in an exponential form is postulated, given by

$$\sigma(t)=e^{-iH(t)}\sigma(0), \tag{2.68}$$

in which $H(t)$ is a continuous function of time, that is not to be confused with the Hamiltonian $\mathcal{H}(t)$.[1] As proposed by Magnus [10], $H(t)$ can be written as a series expansion:

$$H(t) = H^{(0)}(t) + H^{(1)}(t) + H^{(2)}(t) + \cdots . \tag{2.69}$$

For reference, the first two terms are

$$H^{(0)}(t) = \int_0^t \mathcal{H}(t_1)\, dt_1, \tag{2.70}$$

$$H^{(1)}(t) = -\frac{i}{2} \int_0^t \int_0^{t_2} [\mathcal{H}(t_2), \mathcal{H}(t_1)]\, dt_1 dt_2. \tag{2.71}$$

In the case of a periodic Hamiltonian, i.e., $\mathcal{H}(t) = \mathcal{H}(t+\tau)$, one can compute the integrals over one cycle and extend them to arbitrary durations by setting

$$\sigma(N\tau) = \left[e^{-iH(\tau)}\right]^N \sigma(0). \tag{2.72}$$

In the literature, the terms of the expansion of $H(\tau)$ are divided by τ to yield a time-independent effective Hamiltonian. The average Hamiltonian is given by the zeroth-order term

$$\overline{H}^{(0)} = \frac{H^{(0)}(\tau)}{\tau}, \tag{2.73}$$

whereas the first-order correction to $\overline{H}^{(0)}$ is given by

$$\overline{H}^{(1)} = \frac{H^{(1)}(\tau)}{\tau}, \tag{2.74}$$

and so on for higher-order terms.

Although a considerable simplification in the calculations is achieved, one drawback of AHT is that the system can only be observed "stroboscopically" at integer multiples of the period, i.e., at $t = N\tau$. A second, possibly more serious, limitation is that the series in Eq. 2.69 must converge. A rough criterion is developed as follows. $H^{(n)}(t)$ contains n-fold products of $\mathcal{H}(t)$ and each integration introduces a term proportional to τ. Therefore, the nth-order term is roughly proportional to $\left(\langle \mathcal{H}^2 \rangle^{1/2} \tau\right)^n$. Hence, the series should converge for $\langle \mathcal{H}^2 \rangle^{1/2} \tau < 1$. Although often useful, this criterion is not rigorous and can provide misleading results.

[1] Magnus [10] shows that, rigorously speaking, $H(t)$ *always* exists for t close to zero, but may not be a solution valid for the entire domain of the function. Indeed, some restrictions must be placed on $\mathcal{H}(t)$ for $H(t)$ to be well defined for all t. The reader is referred to the original work of Magnus for a discussion of these restrictions.

2.5 Average Liouvillian Theory

The behavior of a spin system under the combined effect of *rf* pulses, free-precession intervals and relaxation can be described, with good accuracy, in the framework of the Average Liouvillian Theory (ALT). We shall present here its basic concepts following the account of [11].

It is possible to write a general solution to Eq. 2.25 in the form

$$\sigma(t) = e^{-\hat{L}t_n} \prod_{j=1}^{n-1} \left[\hat{R}_j e^{-\hat{L}t_j} \right] \sigma(0), \tag{2.75}$$

in which $t = \sum_j t_j$, and $\hat{R}_j$ is a superoperator that corresponds to the transformation induced by a pulse, or a group of pulses, which may or may not include free precession delays. In Eq. 2.75, we implicitly assumed that all the transformations induced by pulses are instantaneous. If this assumption is not valid, the effect of the *rf* pulses may be included in $\hat{L}$. Also, in the symbol $\prod_j$ it is implied that the indices j are sorted by time from right to left in ascending order.

For the jth time step a transformed Liouvillian $\hat{L}_j^T$ can be defined as

$$\hat{L}_j^T = \hat{R}_{n-1}\hat{R}_{n-2} \cdots \hat{R}_j \hat{L} \hat{R}_j^{-1} \cdots \hat{R}_{n-2}^{-1} \hat{R}_{n-1}^{-1}. \tag{2.76}$$

Similarly, a transformed initial density matrix is defined as

$$\sigma^T(0) = \prod_{j=1}^{n-1} \hat{R}_j \sigma(0). \tag{2.77}$$

Equations 2.76 and 2.77 allow to rewrite Eq. 2.75 as

$$\sigma(t) = \prod_{j=1}^{n} e^{-\hat{L}_j^T t_j} \sigma^T(0). \tag{2.78}$$

Imperfections of the transformations $\hat{R}_j$ lead to losses that may be treated including in Eq. 2.78 a scalar factor A_j, which correspond to the attenuation due to the jth step:

$$\begin{aligned} \sigma(t) &= \prod_{j=1}^{n} A_j e^{-\hat{L}_j^T t_j} \sigma^T(0) \\ &= A \prod_{j=1}^{n} e^{-\hat{L}_j^T t_j} \sigma^T(0), \end{aligned} \tag{2.79}$$

in which $A = \prod_{j=1}^{n} A_j$ is a measure of the total attenuation due to the entire pulse sequence, with $0 < A < 1$.

In analogy with the average Hamiltonian of Eq. 2.69, the average Liovillian is defined as

$$\hat{L}_{av} = \hat{L}^{(0)} + \hat{L}^{(1)} + \hat{L}^{(2)} \ldots, \tag{2.80}$$

where

$$\begin{aligned}
\hat{L}^{(0)} &= \frac{1}{t} \sum_j \hat{L}_j^T t_j, \\
\hat{L}^{(1)} &= -\frac{1}{2t} \sum_{j>k} \left[\hat{L}_j^T t_j, \hat{L}_k^T t_k \right], \\
\hat{L}^{(2)} &= -\frac{1}{6t} \sum_{j>k>l} \left[\hat{L}_j^T t_j, \left[\hat{L}_k^T t_k, \hat{L}_l^T t_l \right] \right] + \left[\left[\hat{L}_j^T t_j, \hat{L}_k^T t_k \right], \hat{L}_l^T t_l \right] \\
&\quad + \frac{1}{2} \left\{ \left[\hat{L}_k^T t_k, \left[\hat{L}_k^T t_k, \hat{L}_l^T t_l \right] \right] + \left[\left[\hat{L}_k^T t_k, \hat{L}_l^T t_l \right], \hat{L}_l^T t_l \right] \right\} \cdots
\end{aligned} \tag{2.81}$$

Using the definition of average Liouvillian in Eq. 2.79, we obtain

$$\sigma(t) = A e^{-\hat{L}_{av} t} \sigma^T(0) = A \hat{U}_{av} \sigma^T(0). \tag{2.82}$$

The terms in $\hat{L}_{av}$ have symmetry properties that reflect those of the pulse sequence. Indeed, a handful of theorems can be used to simplify the calculation of the average Liouvillian:

1. In the case of a symmetric Liouvillian, i.e., $\hat{L}_j^T = \hat{L}_{n-j+1}^T$, and all the even-order corrections vanish, i.e., $\hat{L}^{(k)} = 0$ for even k;
2. If the transformations $\hat{R}_j$ are distributed in an antisymmetric way, i.e., $\hat{R}_j = \hat{R}_{n-j}^{-1}$, the Liouvillian is symmetric, i.e., $\hat{L}_j^T = \hat{L}_{n-j+1}^T$;
3. On the contrary, if the distribution of the transformations is symmetric, i.e., $\hat{R}_j = \hat{R}_{n-j}$, the Liouvillian is symmetric only if the matrix associated to each transformation is diagonal.

Similarly to the average Hamiltonian, the existence of the average Liouvillian is guaranteed only if some requirements are met. Formally, a given Liouvillian can be written as the sum of a time-dependent and a time-independent part:

$$\hat{L}(t) = \hat{L}_0 + \lambda \hat{L}_1(t). \tag{2.83}$$

Here, $\hat{L}_1(t)$ includes all the time-dependent components that are induced by the transformations applied to the system,[2] whereas $\hat{L}_0$ includes phenomena that are constant during the experiment, such as relaxation, chemical shifts and scalar couplings.

The most important parameter in the above equation is λ, that is a measure of the strength of the influence of the applied *rf* fields on the natural spin dynamics. Indeed, the basis of ALT is the assumption that the propagator can be written at any time in an exponential form, i.e.,

$$\hat{U}(t) = e^{-\Omega(\lambda,t)}. \tag{2.84}$$

In other words, in analogy with the average Hamiltonian theory, the existence of a valid solution to the master equation that can be expressed in an exponential form is assumed. As shown by Maricq [12], the condition of validity is that $\Omega(\lambda, t)$ can be expanded in a power series in λ, although for some singularities it may not be possible to write an exponential solution to the master equation. The reader is referred to [12] for further details.

Lastly, a criterion to estimate the weight of the contribution of higher-order terms to the average Liouvillian is provided. We shall start by defining the norm $\|A\|$ of a matrix A as the square root of the largest eigenvalue λ_{max} of $(A^{\mathsf{T}}A)$.[3]

The following general expression can be deduced from Eq. 2.81:

$$\hat{L}^{(k)} = -\frac{1}{k}\sum_{j>}\left[\hat{L}_j^T t_j, \hat{L}^{(k-1)}\right], \tag{2.85}$$

in which the subscript $j >$ was used to indicate that the time sorting of j must be conserved. If one assumes that the eigenvalues of $\hat{L}$ are good approximations of the eigenvalues of $\hat{L}_j^T$, then

$$\frac{\|\hat{L}^{(k)}\|}{\|\hat{L}^{(k-1)}\|} \approx \frac{\sqrt{\lambda_{max}\left(\hat{L}^{\mathsf{T}}\hat{L}\right)}\left(\sum_j t_j\right)}{k} = \frac{t\sqrt{\lambda_{max}\left(\hat{L}^{\mathsf{T}}\hat{L}\right)}}{k}. \tag{2.86}$$

Higher-order terms can therefore be ignored when the above ratio is much smaller than one, i.e., if $t\sqrt{\lambda_{max}\left(\hat{L}^{\mathsf{T}}\hat{L}\right)} \ll k$, or simply $t\lambda_{max}\left(\hat{L}\right) \ll k$ in the case of a symmetric Liouvillian. Usually, the terms with $k > 2$ can be safely neglected.

[2] The time dependence is not the one due to the modulation of the *rf* fields that can be removed by transforming to a rotating frame, but the one due to the fact that different transformations $\hat{R}_j$ are applied at different t_j. Therefore, $\hat{L}_1(t)$ is, strictly speaking, a *discrete* function of time.

[3] If A is symmetric, $\|A\|$ is simply equal to its largest eigenvalue.

References

1. Cavanagh J, Fairbrother WJ, Palmer AG, Skelton NJ, Rance M (2006) Protein NMR spectroscopy: principle and practice, 2nd edn. Academic Press, Burlington
2. Bloch F (1946) Phys Rev 70:460
3. Solomon I (1955) Phys Rev 99:559–565
4. Wangsness RK, Bloch F (1953) Phys Rev 89:728–739
5. Redfield AG (1965) Adv Magn Reson 1:1–32
6. Hubbard PS (1969) Phys Rev 180:319–326
7. Abragam A (1961) Principles of Nuclear Magnetism. Clarendon Press, Oxford
8. Maricq MM (1982) Phys Rev B 25:6622
9. Haeberlen U, Waugh JS (1968) Phys Rev 175:453
10. Magnus W (1954) Commun Pure Appl Math 7:649
11. Ghose R (2000) Concept Magnetic Res 12:152–172
12. Maricq MM (1990) Adv Magn Reson 14:151–182

Chapter 3
Analytical Models for Relaxation Dispersion Experiments

In the following, we shall use the theoretical principles discussed in the previous chapter to develop analytical models to describe relaxation dispersion experiments and to derive analytical functions that allow one to extract information from the experimental data. We shall limit our discussion to the case of fast exchange between two conformers, which turned out to be the most useful model to describe our experimental results.

3.1 Exchange Contributions to Relaxation in the Absence of *rf* Fields

The free precession of transverse magnetization can be described in the framework of the McConnell equations (see Eq. 2.22) with $\omega_1 = 0$ [1]. This allows one to quantify contributions of exchange processes to the relaxation rate of interest, i.e., the sensitivity of the relaxation dispersion method of choice, and to determine the asymptotic behavior of the relaxation rate under study in the limit of infinite amplitude of the applied *rf* fields.

3.1.1 *Effects in Single-Quantum Spectroscopy*

The frequency domain spectrum is given by

$$\left\langle \mathbf{M}^{+}(\omega) \right\rangle = \left\langle \psi_0 \right| \left\{ i(\omega - \Omega) + \rho_2 - \mathbf{K}' \right\}^{-1} \left| \psi_0 \right\rangle \mathbf{M}^{+}(0), \tag{3.1}$$

in which $\Omega_{ij} = \delta_{ij}\Omega_i$ contains information about the chemical shifts, $\rho_{2ij} = \delta_{ij} R_{2i}^0$ describes the effects of exchange-independent relaxation mechanisms, and the matrix

N. Salvi, *Dynamic Studies Through Control of Relaxation in NMR Spectroscopy*, Springer Theses, DOI: 10.1007/978-3-319-06170-2_3,

$\mathbf{K}' = \mathbf{SKS}^{-1}$, with $S_{ij} = \delta_{ij} p_i^{-1/2}$, is the symmetrized exchange matrix that takes into account the fact that, at equilibrium, the ith site has fractional population p_i.

Except for very slow exchange, it is a good approximation to replace the R_{2i}^0 by the average value $\bar{R}_2^0$, since $|R_{2i}^0 - \bar{R}_2^0| \ll K_{ii}$.

For fast exchange between two sites A and B, with forward and backward rates k_{AB} and k_{BA}, respectively, when only a single resonance is observed, the transverse relaxation rate constant is [2]

$$R_2 = \bar{R}_2^0 + \frac{\langle u_1|\omega|u_2\rangle^2}{E_2}. \tag{3.2}$$

where $u_{1,2}$ and $E_{1,2}$ are the eigenvectors and the modulus of the eigenvalues of $\mathbf{K}'$. Using the above equation, one obtains the final result [3, 4]

$$R_2 = \bar{R}_2^0 + \frac{p_1 p_2 \Delta\omega^2}{k_{ex}} = \bar{R}_2^0 + R_2^{ex}, \tag{3.3}$$

in which $k_{ex} = k_{AB} + k_{BA}$ and $\Delta\omega$ is the difference between the resonance frequencies of the two sites.

3.1.2 Effects in Multiple-Quantum Spectroscopy

If two spins I and S are affected by the same exchange process as in the previous section, *correlated* transitions of the two nuclei between the sites result in *correlated* chemical shift changes. In this case, chemical exchange can induce either *broadening* or *narrowing* of the resonance line [5].

The chemical shift differences to be used when investigating the relaxation properties of multiple-quantum coherences can be defined as $\Delta\omega_{DQ} = \Delta\omega_I + \Delta\omega_S$ for the double-quantum (DQ) coherence I^+S^+ and $\Delta\omega_{ZQ} = \Delta\omega_I - \Delta\omega_S$ for the zero-quantum (ZQ) coherence I^+S^-.

Using the above expressions in Eq. 3.3, the difference in relaxation rate constants for the multiple-quantum coherences is given by

$$\Delta R_{MQ} = R_{DQ} - R_{ZQ} = \Delta R_{MQ}^0 + \frac{4 p_1 p_2 \Delta\omega_I \Delta\omega_S}{k_{ex}} = \Delta R_{MQ}^0 + \Delta R_{MQ}^{ex}, \tag{3.4}$$

with $\Delta R_{MQ}^0 = R_{DQ}^0 - R_{ZQ}^0$. This exchange-free rate, being the difference of two rates of the same order of magnitude, is very close to zero. Thus, a significant value for ΔR_{MQ} is a clear indication of the presence of an exchange process. Also, the sign of ΔR_{MQ} is essentially determined by the relative sign of the chemical shift differences $\Delta\omega_I$ and $\Delta\omega_S$. This information, which is not provided by single-quantum experiments, can offer insights into the mechanism of the exchange phenomenon under investigation [6–8].

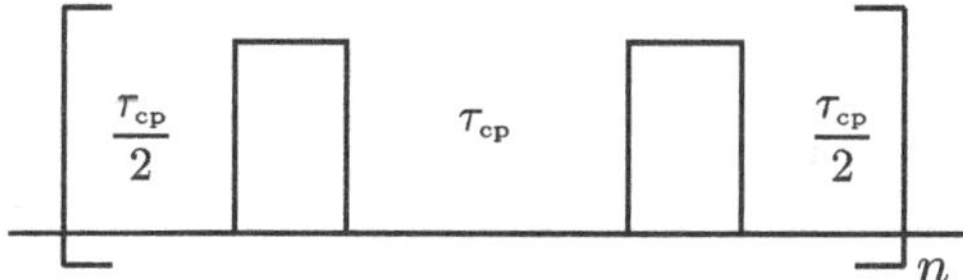

Fig. 3.1 Basic CPMG echo train. During a CPMG experiment, a train of π pulses (*empty rectangles*) is applied. The spacing between consecutive pulses is τ_{cp}. The block $\tau_{cp}/2 - \pi$ pulse$-\tau_{cp} - \pi$ pulse$-\tau_{cp}/2$ is repeated n times

3.2 Relaxation During CPMG Echo Trains

3.2.1 Relaxation of Single-Quantum Coherences

In single-quantum CPMG experiments the relaxation of transverse magnetization is observed during a train of π pulses (see Fig. 3.1).

The effect of each of these pulses is to invert the sense of precession of the magnetization. According to [2], the transverse relaxation rate is given by

$$R_2\left(1/\tau_{cp}\right) = \bar{R}_2^0 - \frac{1}{2\tau_{cp}} \ln \lambda, \tag{3.5}$$

in which λ is the largest eigenvalue of the matrix $\mathbf{A} = \exp\left[(i\omega + \Gamma)^{\dagger}\,\tau_{cp}\right] \exp\left[(i\omega + \Gamma)\,\tau_{cp}\right]$.

In the case of two sites in fast exchange, Eq. 3.5 simplifies to

$$R_2\left(1/\tau_{cp}\right) = \bar{R}_2^0 + \frac{p_1 p_2 \Delta\omega^2}{k_{ex}}\left(1 - \frac{2\tanh\left[k_{ex}\tau_{cp}/2\right]}{k_{ex}\tau_{cp}}\right). \tag{3.6}$$

3.2.2 Relaxation of Multiple-Quantum Coherences

In multiple-quantum CPMG experiments it is convenient to select either the zero- or double-quantum component and to analyze the respective data separately. Therefore, without loss of generality, it is possible to consider only the zero-quantum coherences in the theoretical framework offered by the McConnell equations. Assuming exchange between two sites A and B, under the effect of a spin-echo applied to the two channels simultaneously (see Fig. 3.2) and on resonance with the resonance frequency of the two spins in site A, the zero-quantum coherences will evolve as follows:

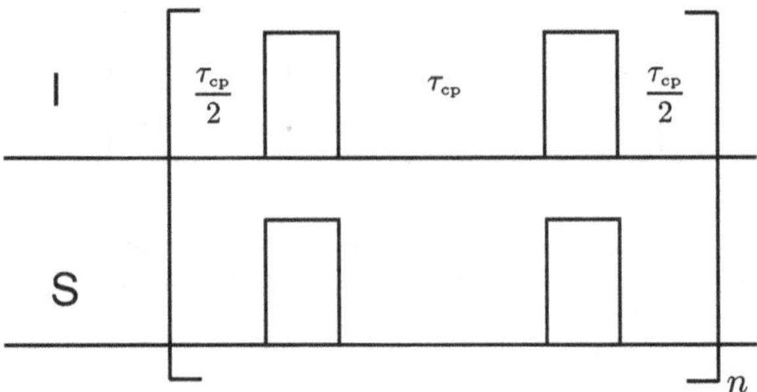

Fig. 3.2 Basic multiple-quantum CPMG echo train. During a multiple-quantum CPMG experiment, a train of π pulses (*empty rectangles*) is applied simultaneously on the two channels I and S. The spacing between consecutive pulses is τ_{cp}. The block $[\tau_{cp}/2 - \pi$ pulse$-\tau_{cp} - \pi$ pulse$-\tau_{cp}/2]$ is repeated n times

$$\begin{bmatrix} ZQ_x^A(1/2\nu_{cp}) \\ ZQ_x^B(1/2\nu_{cp}) \\ ZQ_y^A(1/2\nu_{cp}) \\ ZQ_y^B(1/2\nu_{cp}) \end{bmatrix} = e^{\mathbf{A}/4\nu_{cp}} R^{\pi} e^{\mathbf{A}/4\nu_{cp}} \begin{bmatrix} ZQ_x^A(0) \\ ZQ_x^B(0) \\ ZQ_y^A(0) \\ ZQ_y^B(0) \end{bmatrix} = e^{\mathbf{A}/4\nu_{cp}} R^{\pi} e^{\mathbf{A}/4\nu_{cp}} \begin{bmatrix} p_A \\ 1-p_A \\ 0 \\ 0 \end{bmatrix}, \tag{3.7}$$

in which the matrix

$$\mathbf{A} = \begin{bmatrix} -k_{ex}(1-p_A) - R_{ZQ}^0 & k_{ex}p_A & 0 & 0 \\ k_{ex}(1-p_A) & -k_{ex}p_A - R_{ZQ}^0 & 0 & -\Delta\omega_{ZQ} \\ 0 & 0 & -k_{ex}(1-p_A) - R_{ZQ}^0 & k_{ex}p_A \\ 0 & \Delta\omega_{ZQ} & k_{ex}(1-p_A) & -k_{ex}(1-p_A) - R_{ZQ}^0 \end{bmatrix} \tag{3.8}$$

includes both the effects of auto-relaxation and of chemical exchange, p_A is the population of site A, R_{ZQ}^0 is the exchange-free auto-relaxation of zero-quantum coherences, $\Delta\omega_{ZQ} = \Delta\omega_I - \Delta\omega_S$ is the difference in chemical shift between the two states revelant for the zero-quantum coherences, and

$$R^{\pi} = \begin{bmatrix} 1 & 0 & 0 & 0 \\ 0 & 1 & 0 & 0 \\ 0 & 0 & -1 & 0 \\ 0 & 0 & 0 & -1 \end{bmatrix} \tag{3.9}$$

is a matrix representing the effect of an ideal π pulse with phase x. Thus, the relaxation rate of zero-quantum coherences can be expressed as

$$R_{ZQ}(\nu_{cp}) = -\frac{1}{2\nu_{cp}} \log\left[\frac{ZQ_x^A(1/2\nu_{cp})}{ZQ_x^A(0)}\right] = -\frac{1}{2\nu_{cp}} \log\left[\frac{ZQ_x^A(1/2\nu_{cp})}{p_A}\right]. \tag{3.10}$$

Through cumbersome though elementary algebraic calculations, the above equation can be rewritten as

$$R_{\mathrm{ZQ}}\left(\nu_{\mathrm{cp}}\right) = \mathrm{Re}\left\{-2\nu_{\mathrm{cp}}\log\left[e^{-\frac{k_{ex}}{4\nu_{\mathrm{cp}}}}\frac{CD}{AB}\right]\right\} + R^{0}_{\mathrm{ZQ}}, \tag{3.11}$$

in which

$$\begin{aligned} A &= \sqrt{k_{ex}^2 - 2ik_{ex}\left(-1+2p_{\mathrm{A}}\right)\Delta\omega_{\mathrm{ZQ}} - \Delta\omega_{\mathrm{ZQ}}^2} \\ B &= \sqrt{k_{ex}^2 + 2ik_{ex}\left(-1+2p_{\mathrm{A}}\right)\Delta\omega_{\mathrm{ZQ}} - \Delta\omega_{\mathrm{ZQ}}^2} \\ C &= A\cosh\left[\frac{A}{8\nu_{\mathrm{cp}}}\right] + \left(k_{ex} - i\Delta\omega_{\mathrm{ZQ}}\right)\sinh\left[\frac{A}{8\nu_{\mathrm{cp}}}\right] \\ D &= B\cosh\left[\frac{B}{8\nu_{\mathrm{cp}}}\right] + \left(k_{ex} + i\Delta\omega_{\mathrm{ZQ}}\right)\sinh\left[\frac{B}{8\nu_{\mathrm{cp}}}\right]. \end{aligned} \tag{3.12}$$

An analogous expression can be derived using the same procedure for the relaxation of double-quantum coherences. In this case, $\Delta\omega_{\mathrm{ZQ}}$ is replaced by $\Delta\omega_{\mathrm{DQ}} = \Delta\omega_{\mathrm{I}}+\Delta\omega_{\mathrm{S}}$.

Expressions similar to those derived in this sections have been published in [9].

3.3 Relaxation During $\mathbf{R_{1\rho}}$ Experiments

In $\mathrm{R}_{1\rho}$ experiments the relaxation rate of single-quantum coherences is measured while applying a continuous-wave *rf* field at $\omega_{\mathrm{carrier}}$ with amplitude $\omega_1/(2\pi)$. The auto-relaxation rate for the component of the magnetization along the direction of the effective field, which is a function of the tilt angle θ (see Eq. 2.55), depends on ω_1 and on the population-averaged chemical shift Ω. More precisely, $\mathrm{R}_{1\rho}$ depends on the effective field amplitude in the rotating frame $\omega_{\mathrm{eff}} = \sqrt{\omega_1^2 + \Omega^2}$.

Let us assume an exchange process between two sites, in which ω_{eff} is the same for both sites. In other words, we are assuming that either ω_1 or the offsets of both sites are much larger than the chemical shift difference between the two sites $\Delta\omega$. Under this assumption, the following expression is valid in the fast-exchange regime [10, 11]:

$$R_{1\rho} = R^{0}_{1\rho} + \sin^2\theta\frac{p_{\mathrm{A}}p_{\mathrm{B}}\Delta\omega^2 k_{ex}}{k_{ex}^2 + \omega_{\mathrm{eff}}^2}, \tag{3.13}$$

in which $R^{0}_{1\rho}$ is the exchange-free relaxation rate constant, i.e. the relaxation rate constant in the limit $\omega_{\mathrm{eff}} \to \infty$. Combining Eqs. 2.55 and 3.13, one obtains the following expression for the relaxation of the transverse component of the magnetization:

$$R_2 = R_2^0 + \frac{p_\mathrm{A} p_\mathrm{B} \Delta\omega^2 k_{ex}}{k_{ex}^2 + \omega_\mathrm{eff}^2}, \tag{3.14}$$

where R_2^0 is the exchange-free relaxation rate constant. The above expression allows one to extract from the relaxation dispersion information about the product $p_\mathrm{A} p_\mathrm{B} \Delta\omega^2$ of the populations and the square of the chemical shift difference between the exchanging sites, but does not allow one to determine these parameters separately.

3.4 Evolution and Relaxation of Multiple-Quantum Coherences Under Heteronuclear Double Resonance Irradiation

3.4.1 Relationship Between ΔR_{MQ} and the Cross-Relaxation Rate of Multiple-Quantum Coherences

In order to perform relaxation dispersion experiments of multiple-quantum coherences and extract information from the value of ΔR_{MQ} as shown in Sect. 3.1.2, it is possible to measure the cross-relaxation rate of multiple-quantum coherences μ_MQ, defined as the rate of the following interconversion process:

$$2I_x S_x \longleftrightarrow 2I_y S_y. \tag{3.15}$$

Indeed, in the absence of any applied *rf* fields, the effect of the relaxation superoperator on the evolution of multiple-quantum coherences is described by a set of differential equations as follows:

$$\frac{\mathrm{d}}{\mathrm{d}t}\begin{bmatrix} 2I_x S_x \\ 2I_y S_y \end{bmatrix} = \begin{bmatrix} -\lambda_\mathrm{MQ} & \mu_\mathrm{MQ} \\ \mu_\mathrm{MQ} & -\lambda_\mathrm{MQ} \end{bmatrix}\begin{bmatrix} 2I_x S_x \\ 2I_y S_y \end{bmatrix}, \tag{3.16}$$

in which λ_MQ is the auto-relaxation rate.

If we start with $\rho(0) = 2I_x S_x$, the solution of the above system of coupled equations is given by

$$\begin{aligned} 2I_x S_x\,(t) &= e^{-(\lambda_\mathrm{MQ}+\mu_\mathrm{MQ})t} + e^{-(\lambda_\mathrm{MQ}-\mu_\mathrm{MQ})t}, \\ 2I_y S_y\,(t) &= e^{-(\lambda_\mathrm{MQ}-\mu_\mathrm{MQ})t} - e^{-(\lambda_\mathrm{MQ}+\mu_\mathrm{MQ})t}. \end{aligned} \tag{3.17}$$

Recalling the definitions

$$\begin{aligned} 2I_x S_x &= ZQ_x + DQ_x, \\ 2I_y S_y &= ZQ_x - DQ_x, \end{aligned} \tag{3.18}$$

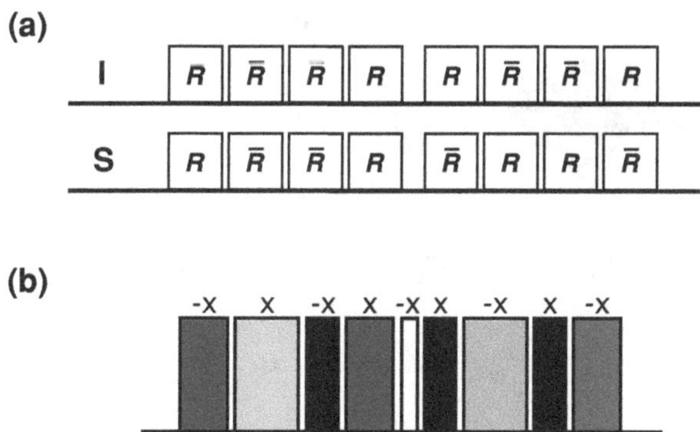

Fig. 3.3 HDR WALTZ-32 pulse scheme. **a** The first part of the sequence corresponds to a WALTZ-16 sequence applied simultaneously, with the same phases and *rf* amplitudes, to both I and S. During the second part of the sequence the phase of the composite pulses for spin S are shifted by π. **b** The scheme is based on the repetition of a composite pulse R shown here. 90, 180, 270 and 360° pulses are represented by *white*, *black*, *dark grey* and *light gray rectangles*, respectively. The labels at the *top* of each *rectangle* indicate the phases of the *rf* pulses

it is straightforward to recognize from Eq. 3.17 that

$$\begin{aligned} R_{\text{ZQ}} &= \lambda_{\text{MQ}} - \mu_{\text{MQ}}, \\ R_{\text{DQ}} &= \lambda_{\text{MQ}} + \mu_{\text{MQ}}. \end{aligned} \tag{3.19}$$

Therefore, we can conclude that

$$\Delta R_{\text{MQ}} = R_{\text{DQ}} - R_{\text{ZQ}} = 2\mu_{\text{MQ}}. \tag{3.20}$$

3.4.2 Preservation of Relaxation Pathways

However, the simultaneous application of continuous-wave *rf* fields to both I and S, in a fashion analogous to $R_{1\rho}$ experiments, would not preserve the interconversion between the multiple-quantum coherences. For instance, by applying the fields along the x axes of the the doubly rotating frame, one would effectively spin-lock $2I_xS_x$, wheres $2I_yS_y$ would nutate about the *rf* fields and it would be dephased by *rf* inhomogeneities.

Therefore, the simple continuous-wave spin-lock has to be replaced by a pulse scheme able to preserve all multiple-quantum coherences in the relevant subspace (i.e., $2I_xS_x$, $2I_yS_y$, $2I_xS_y$, $2I_yS_x$), so that cross-relaxation is the only process that can lead to an interconversion between these operators.

The WALTZ-32 scheme (see Fig. 3.3) has the required properties [18]. Indeed, using the formalism of the Average Hamiltonian Theory presented in the previous chapter, it is possible to show that

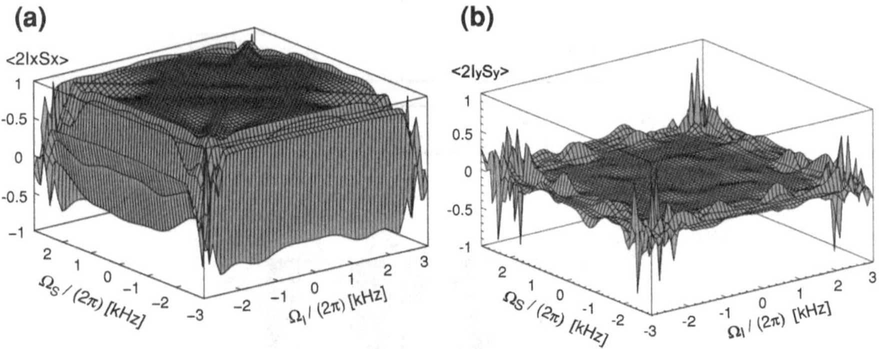

Fig. 3.4 Reprinted with permission from [18]. Copyright 2009, AIP Publishing LLC. Simulations of the expectation values of $2I_x S_x$ (**a**) and $2I_y S_y$ (**b**) after two cycles of HDR WALTZ-32 irradiation as a function of the offsets of spins I and S. The initial density operator was set to $2I_x S_x$, with a scalar coupling constant J = 90 Hz and an *rf* amplitude $\omega_1 / (2\pi) = 2\,\text{kHz}$

$$\bar{H}^{(0)}_{\text{WALTZ}} = \frac{\pi J_{\text{IS}}}{2} 2I_z S_z, \tag{3.21}$$

and

$$\bar{H}^{(2)}_{\text{WALTZ}} = -\frac{1}{\omega_1^2}\frac{J_{\text{IS}}}{576}\left[21 J_{\text{IS}}^2 \pi^3 + 4\,(10 + 49\pi)\left(\Omega_{\text{I}}^2 + \Omega_{\text{S}}^2\right)\right] 2I_z S_z. \tag{3.22}$$

Thus, all the orders of the average Hamiltonian computed by [18] commute with the multiple-quantum operators so that offsets and scalar couplings can be ignored, provided that their magnitude is smaller than the applied *rf* pulses.

These results seem to be confirmed by the output of numerical simulations, where the expectation values of the multiple-quantum coherences of interest under HDR WALTZ-32 for different values of the offsets of spins I and S are determined (see Fig. 3.4). The offset profiles show that during the sequence the coherent leakage between $2I_x S_x$ and $2I_y S_y$ becomes significant only for offsets larger than ω_1.

However, our experimental results do not confirm the output of numerical simulations. Indeed, as shown in Fig. 3.5, significant leakage from $2I_x S_x$ is observed also for offsets smaller than the *rf* amplitude, and even for very small offsets. The leakage measured when using older electronics (Fig. 3.5a) is generally larger than the one measured when the experiments are carried out using newer equipment (Fig. 3.5b). Also, the results are only partially reproducible when using spectrometers of the same generation. These observations support the hypothesis that the discrepancy between theory and experiments is due to pulse imperfections and in particular to the fact that a finite amount of time is required to switch the phase of the *rf* pulses from x to −x and vice-versa, whereas a windowless sequence with instantaneous phase switching is assumed in the calculations. Nevertheless, attempts to reproduce the experimental errors using simple modifications of our numerical simulations, for

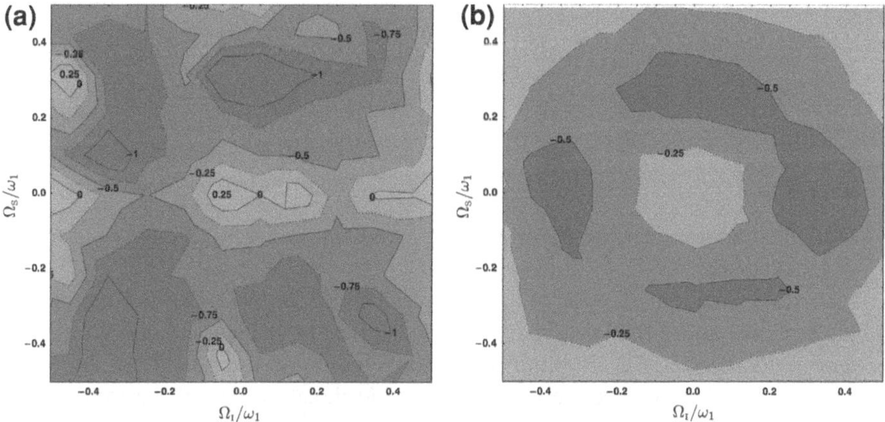

Fig. 3.5 Measurements of the expectation values of $2I_xS_x$ **a** at 11.7 T (500 MHz for ^{1}H) using a Bruker Avance II console for the generation of the *rf* pulses and **b** at 14.1 T (600 MHz for ^{1}H) using a Bruker Avance III console for the generation of the *rf* pulses as a function of the offsets of spins I and S. A state $2I_xS_x$ was prepared in the H^N-N pair of the side chain of L-tryptophan enriched in ^{15}N dissolved in a mixture of 90 % ethylene glycol and 10 % H_2O. D_2O was used as an external lock in an inner concentric tube. The HDR irradiation was applied during three cycles of WALTZ-32 with *rf* amplitudes $\omega_1/(2\pi) = 1$ kHz on both channels. Contours delimitate areas where the measured values are ±0.25, ±0.50, ±0.75 and ±1.0 times the maximum measured value

instance by including delays between the pulses, were not successful. For the time being reliable data can only be obtained using the HDR technique when the *rf* pulses are applied on resonance for both spins I and S.

3.4.3 Effect of the HDR Irradiation on Relaxation

The effect of the HDR WALTZ-32 sequence of the previous section can be described in the framework of average Liouvillian theory [16]. In the following it will be assumed that the system is observed stroboscopically at integer multiples of the period $T = 96\pi/\omega_1$. On the basis of the results of average Hamiltonian theory, scalar couplings and offsets were not included in the calculations. Using Eq. 2.82, and considering only zero-th order effects, one obtains

$$\frac{\mathrm{d}}{\mathrm{d}t}\begin{bmatrix} 2I_xS_x \\ 2I_yS_y \end{bmatrix} = \begin{bmatrix} -\lambda_{\mathrm{xx}}^{\mathrm{WALTZ}} & \mu_{\mathrm{MQ}}^{\mathrm{WALTZ}} \\ \mu_{\mathrm{MQ}}^{\mathrm{WALTZ}} & -\lambda_{\mathrm{yy}}^{\mathrm{WALTZ}} \end{bmatrix}\begin{bmatrix} 2I_xS_x \\ 2I_yS_y \end{bmatrix}, \tag{3.23}$$

in which

$$\begin{aligned} \lambda_{xx}^{\text{WALTZ}} &= \lambda_{\text{MQ}}, \\ \lambda_{yy}^{\text{WALTZ}} &= \frac{1}{8}\left(3\lambda_{\text{MQ}} + \rho_{\text{I}}^{\text{a}} + \rho_{\text{S}}^{\text{a}} + 3\rho_{\text{IS}}\right), \\ \mu_{\text{MQ}}^{\text{WALTZ}} &= \mu_{\text{MQ}}/2. \end{aligned} \tag{3.24}$$

In other words,

1. the effective cross-relaxation rate $\mu_{\text{MQ}}^{\text{WALTZ}}$ is half of the rate in the absence of any *rf* irradiation;
2. the auto-relaxation of $2I_xS_x$ is unaffected by the HDR WALTZ-32 block;
3. the effective auto-relaxation rate of $2I_yS_y$ is given by a linear combination of the auto-relaxation rate λ_{MQ}, the auto-relaxation rates of the antiphase terms $2I_xS_z$ and $2I_zS_x$ ($\rho_{\text{I}}^{\text{a}}$ and $\rho_{\text{S}}^{\text{a}}$, respectively), and the auto-relaxation rate of the longitudinal two-spin order $2I_zS_z$;
4. on the basis of the above equations, it is possible to conclude that the evolution of the multiple-quantum coherences is restricted to a seven-dimensional subspace of the Liouville space, spanned by the operators $2I_xS_x$, $2I_yS_y$, $2I_zS_z$, $2I_zS_x$, $2I_xS_z$, $2I_zS_y$, $2I_yS_z$;
5. the fact that the auto-relaxation rates $\lambda_{xx}^{\text{WALTZ}}$ and $\lambda_{yy}^{\text{WALTZ}}$ are not the same requires the use of *symmetrical reconversion* in cross-relaxation experiments [17]. According to this procedure, four signals $s_{\text{a/b}}$ (a, b = xx, yy) are recorded, by selecting either $2I_xS_x$ or $2I_yS_y$ before and after the HDR irradiation. The signals are then combined to give

$$S(nT) = \sqrt{\frac{s_{\text{xx/yy}}s_{\text{yy/xx}}}{s_{\text{xx/xx}}s_{\text{yy/yy}}}}, \tag{3.25}$$

in which n is an integer. The cross-relaxation rate of interest can than be extracted by fitting the results of several experiments with different n to

$$f(nT) = \tanh|\mu_{\text{MQ}}^{\text{WALTZ}} nT| = \tanh|\frac{\mu_{\text{MQ}}}{2} nT|. \tag{3.26}$$

The use of symmetrical reconversion masks the information about the sign of $\mu_{\text{MQ}}^{\text{WALTZ}}$, which can be recovered by comparing the relative signs of $s_{\text{xx/xx}}$ ($s_{\text{yy/yy}}$) and $s_{\text{xx/yy}}$ ($s_{\text{yy/xx}}$).

3.4.4 Relaxation of Multiple-Quantum Coherences Under HDR Irradiation

Recently, Podkorytov and Skrynnikov [12] have discussed relaxation-dispersion experiments employing a single-resonance spin-lock *rf* field with alternating phases, $[(x)(-x)]_n$, applied to a spin S such as ^{1}H or ^{15}N, and have presented a theoretical model based on Redfield theory that leads to an analytical expression to describe the dispersion, i.e., the dependence on the *rf* amplitude ω_1 of exchange-induced contributions to single-quantum relaxation. In the present section we discuss the more challenging case of cross-relaxation of MQ coherences under the WALTZ-32 HDR irradiation scheme on the basis of Redfield theory [13].

Let us consider a scalar-coupled heteronuclear two-spin system, I and $S = 1/2$, and let us assume that the spin system experiences correlated modulations of the isotropic chemical shifts between two non-equivalent sites or conformations, a and b. These modulations are assumed to be fast compared to the chemical shift differences, i.e., $\Delta\omega_{ab}^{I,S}\tau_{\text{ex}} \ll 1$, where $\tau_{\text{ex}} = 1/k_{\text{ex}} = 1/(k_{\text{f}} + k_{\text{r}})$, k_{f} and k_{r} are the forward and reverse exchange rates, respectively, and $\Delta\omega_{ab}^{k} = \omega_b^k - \omega_a^k$ are the chemical shift differences between sites a and b for spin $k = I, S$. In the fast exchange regime, one observes averaged spectral lines in the single-quantum spectra of spins $k = I, S$ at frequencies $\omega_k = p_a\omega_a^k + p_b\omega_b^k$, in which p_a and p_b represent the populations of sites a and b.

In a doubly-rotating frame (DRF) precessing about the z axis at the two angular frequencies ω_I and ω_S, the exchange-modulated chemical-shift Hamiltonian can be written as

$$H_{\text{z}}(t) = n(t)h_{\text{z}} = n(t)\left[\Delta\omega_{ab}^{I}I_z + \Delta\omega_{ab}^{S}S_z\right], \tag{3.27}$$

in which $n(t)$ is the stochastic function defined by Podkorytov and Skrynnikov [12], i.e., a step-like random function that takes either of two values p_b and $-p_a$ with probabilities p_a and p_b, respectively (see Fig. 3.6). We shall assume here without loss of generality that the chemical shift modulations are fully correlated, i.e., the function $n(t)$ is the same for spins I and S.

In the DRF the Hamiltonian describing the on-resonance *rf* irradiation is

$$H_{\text{rf}}(t) = \omega_1^I(t)I_x + \omega_1^S(t)S_x, \tag{3.28}$$

in which the amplitudes $\omega_1^k(t)$ $(k = I, S)$ take the values ω_1 and $-\omega_1$ if the *rf* phases are switched between $+x$ and $-x$, as shown in Fig. 3.3. In the DRF the dynamics of a MQ coherence, described by the density operator $\sigma_{\text{MQ}}(t)$, in the presence of time-dependent double-resonance *rf* fields is governed by the following equation [12, 14]:

$$\frac{d}{dt}\sigma_{\text{MQ}}(t) = -\int_0^\infty \mathbb{C}(\tau)\left[h_{\text{z}}, \left[e^{-i(\varphi_I(t,\tau)I_x+\varphi_S(t,\tau)S_x)}h_{\text{z}}e^{-i(\varphi_I(t,\tau)I_x+\varphi_S(t,\tau)S_x)}, \sigma_{\text{MQ}}(t)\right]\right]d\tau, \tag{3.29}$$

with the correlation function

$$\mathbb{C}(\tau) = \langle n(t)n(t+\tau)\rangle = p_a p_b e^{-|\tau|/\tau_{\text{ex}}}, \tag{3.30}$$

and the phase

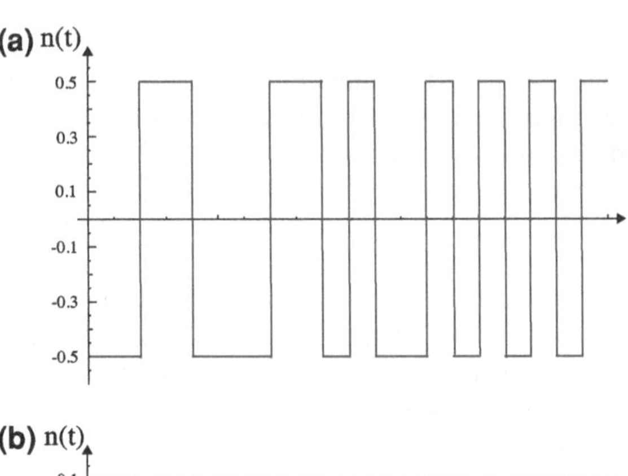

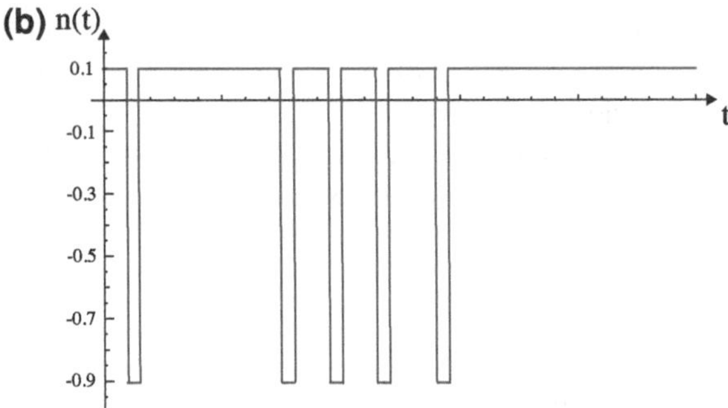

Fig. 3.6 Graphical representation of the stochastic function $n(t)$ for the case of (**a**) equal populations $p_a = p_b = 0.5$ and (**b**) asymmetric populations $p_a = 0.9$ and $p_b = 0.1$

$$\varphi_k(t, \tau) = \int_{t-\tau}^{t} \omega_1^k(t')dt' \qquad (k = I, S). \tag{3.31}$$

Equation 3.29 describes the relaxation of σ_{MQ} due to chemical-exchange-induced fluctuations of the isotropic chemical shifts under heteronuclear double-resonance *rf* irradiation. Information about the stochastic modulations of the chemical shifts is contained in the time dependence of $n(t)$, which appears in the correlation function $\mathbb{C}(\tau)$. Information about the *rf* fields is included in the phases $\varphi_{I,S}(t, \tau)$. Equation 3.29 does not include the coherent evolution of the density operator $\sigma_{\text{MQ}}(t)$ under HDR irradiation, because, as discussed in the previous sections, coherent effects are averaged to zero provided that stroboscopic observation of the spin system after integer numbers of WALTZ-32 cycles is carried out.

The MQ cross-relaxation rate μ_{MQ} that governs the interconversion between MQ operators $\mathbb{X} = 2I_xS_x$ and $\mathbb{Y} = 2I_yS_y$ during *rf* irradiation can be calculated from Eq. 3.29,

$$
\begin{aligned}
\mu_{\text{MQ}}(t) = {} & p_a p_b \left(\Delta\omega_{ab}^I\right)^2 \int_0^\infty e^{-\tau/\tau_{\text{ex}}} \mathbf{Tr}\left\{\mathbb{Y}\left[I_z, \left[e^{-i\varphi_I(t,\tau)I_x} I_z e^{i\varphi_I(t,\tau)I_x}, \mathbb{X}\right]\right]\right\} d\tau \\
& + p_a p_b \left(\Delta\omega_{ab}^S\right)^2 \int_0^\infty e^{-\tau/\tau_{\text{ex}}} \mathbf{Tr}\left\{\mathbb{Y}\left[S_z, \left[e^{-i\varphi_S(t,\tau)S_x} S_z e^{i\varphi_S(t,\tau)S_x}, \mathbb{X}\right]\right]\right\} d\tau \\
& + p_a p_b \Delta\omega_{ab}^I \Delta\omega_{ab}^S \int_0^\infty e^{-\tau/\tau_{\text{ex}}} \mathbf{Tr}\left\{\mathbb{Y}\left[S_z, \left[e^{-i\varphi_I(t,\tau)I_x} I_z e^{i\varphi_I(t,\tau)I_x}, \mathbb{X}\right]\right]\right\} d\tau \\
& + p_a p_b \Delta\omega_{ab}^I \Delta\omega_{ab}^S \int_0^\infty e^{-\tau/\tau_{\text{ex}}} \mathbf{Tr}\left\{\mathbb{Y}\left[I_z, \left[e^{-i\varphi_S(t,\tau)S_x} S_z e^{i\varphi_S(t,\tau)S_x}, \mathbb{X}\right]\right]\right\} d\tau,
\end{aligned}
\tag{3.32}
$$

where we have used the definition $\langle A|B\rangle = \mathbf{Tr}\left\{A^\dagger B\right\}$. One can verify that the first two auto-correlated terms in Eq. (3.32) cannot contribute to the multiple-quantum cross-relaxation rate of interest [15] and that the rate $\mu_{\text{MQ}}(t)$ thus entirely depends on the cross-correlated contributions of the last two terms. Explicit spin-algebra calculations lead to

$$
\mu_{\text{MQ}}(t) = \overline{\Delta\omega}^2 \int_0^\infty e^{-\tau/\tau_{\text{ex}}} \left[\cos\varphi_I(t,\tau) + \cos\varphi_S(t,\tau)\right] d\tau, \tag{3.33}
$$

where $\overline{\Delta\omega}^2 = p_a p_b \Delta\omega_{ab}^I \Delta\omega_{ab}^S$.

The scaling effect of the multiple-quantum cross-relaxation rate during the HDR WALTZ-32 sequence, discussed in the previous section in the framework of the average Liouvillian theory, is not taken into account by Eq. 3.29 as written in the DRF. This averaging effect is included here by introducing *ad hoc* a factor of 1/2. We thus rewrite Eq. 3.33 as

$$
\mu_{\text{MQ}}^{\text{WALTZ}}(t) = \frac{1}{2}\overline{\Delta\omega}^2 \sum_{n=0}^{\infty} j_n(t), \tag{3.34}
$$

in which

$$
j_n(t) = \int_{8n\tau_{\text{R}}}^{8(n+1)\tau_{\text{R}}} e^{-\tau/\tau_{\text{ex}}} \left[\cos\varphi_I(t,\tau) + \cos\varphi_S(t,\tau)\right] d\tau, \tag{3.35}
$$

where $\tau_{\text{R}} = 12\pi/\omega_1$ is the length of one of the composite pulses R of the HDR WALTZ-32 pulse sequence, as presented in Fig. 3.3. Each integration interval of length $8\tau_{\text{R}}$ in Eq. 3.35, $[8n\tau_{\text{R}}, 8(n+1)\tau_{\text{R}}]$, corresponds to the duration of a full

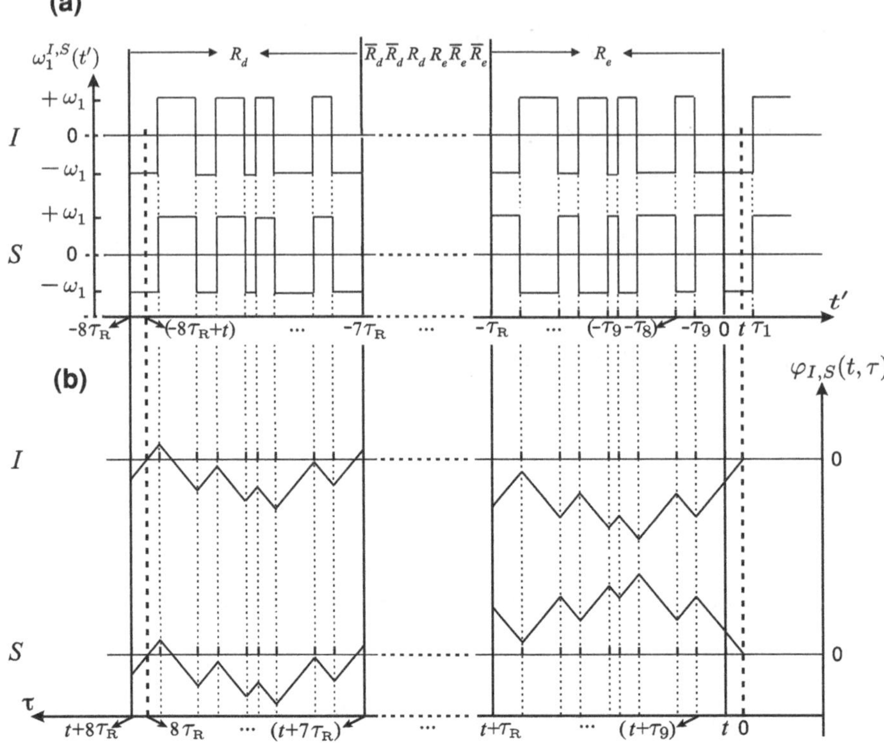

Fig. 3.7 Copyright © 2011 WILEY-VCH Verlag GmbH & Co. KGaA, Weinheim. Schematic representation of **a** $\omega_1^{I,S}(t')$ and **b** $\varphi_{I,S}(t,\tau)$. We used the nomenclature of [18] for the phase of the pulses: R_d corresponds to a composite pulse R applied to both I and S; R_e corresponds to a pulse R applied to I simultaneously with a pulse $\overline{R}$ applied to S; $\overline{R}_d$ and $\overline{R}_e$ correspond respectively to R_d and R_e with inverted phases. We show only the first and the last blocks, R_d and R_e, respectively, of the HDR WALTZ-32 scheme. The lengths of the pulses in a block R_d or R_e expressed in multiples of $\tau_{\pi/2}$, the duration of a $\pi/2$ pulse, are 3, 4, 2, 3, 1, 2, 4, 2 and 3. Blocks R_d and R_e are delimited by vertical solid lines. The thick vertical dashed lines define a cycle of duration $8\tau_R$. The rightmost of these lines corresponds to a time point t within the first pulse of an HDR WALTZ-32 sequence, as explained in the text

WALTZ-32 block. The function $\varphi_k(t,\tau)$ $(k = I, S)$ has a period $8\tau_R$ with respect to τ (see Fig. 3.7). This yields $j_{n+1}(t) = \exp\left[-8\tau_R/\tau_{ex}\right] j_n(t)$.

Therefore one obtains from Eq. 3.34

$$\mu_{MQ}^{WALTZ}(t) = \frac{1}{2}\overline{\Delta\omega}^2 j_0(t)\sum_{n=0}^{\infty} e^{-8n\tau_R/\tau_{ex}} = \frac{1}{2}\overline{\Delta\omega}^2 \frac{j_0(t)}{1 - e^{-8\tau_R/\tau_{ex}}}. \tag{3.36}$$

In this way the multiple-quantum cross-relaxation rate is calculated under the assumption that the *rf* irradiation preceding the instant t is infinitely long. The remaining challenge is the evaluation of the function $j_0(t)$. The procedure outlined by Podkorytov and Skrynnikov [12] for single-quantum coherences becomes much more complex for multiple-quantum coherences because of the complexity of the HDR WALTZ-32 sequence, which comprises 72 pulses of different durations and phases.

Henceforth we shall use the notation τ_i, with $i = 1, 2, \ldots, 9$, to denote the length of the ith pulse within an element R of the HDR WALTZ-32 sequence (see Fig. 3.7). Let us consider a time point t with $0 \le t \le \tau_1$ within the first pulse of the WALTZ-32 block. In practice, we can split $j_0(t)$ in Eq. 3.35 with $n = 0$ into a sum of 73 integrals (see Fig. 3.7b):

$$j_0^{(1)}(t) = \int_0^t G(t,\tau)d\tau + \int_t^{t+\tau_9} G(t,\tau)d\tau + \int_{t+\tau_9}^{t+\tau_9+\tau_8} G(t,\tau)d\tau + \cdots + \int_{t+\tau_9+\tau_8+\cdots+\tau_2}^{t+\tau_R} G(t,\tau)d\tau + \cdots + \int_{t+7\tau_R+\tau_9+\tau_8+\cdots+\tau_2}^{8\tau_R} G(t,\tau)d\tau, \quad (3.37)$$

in which the superscript in $j_0^{(1)}(t)$ refers to the fact that we are considering a time t during the first pulse of the *rf* sequence and $G(t,\tau) = e^{-\tau/\tau_{\mathrm{ex}}}[\cos\varphi_I(t,\tau) + \cos\varphi_S(t,\tau)]$. Within each of the 73 integration intervals, the *rf* phases $\varphi_k(t,\tau)$ ($k = I, S$), defined by Eq. 3.31, are constant. Therefore, the functions $\cos\varphi_I(t,\tau) + \cos\varphi_S(t,\tau)$ in Eq. 3.37 can be written as

$$\cos\left[\int_{t-\tau}^{t}(-\omega_1)dt'\right] + \cos\left[\int_{t-\tau}^{t}(-\omega_1)dt'\right] = 2\cos[\omega_1\tau] \quad \text{for} \quad \tau \in [0,t], \quad (3.38)$$

$$\cos\left[\int_0^t(-\omega_1)dt' + \int_{t-\tau}^{0}(-\omega_1)dt'\right] + \cos\left[\int_0^t(-\omega_1)dt' + \int_{t-\tau}^{0}\omega_1 dt'\right] = \cos[\omega_1\tau] + \cos[\omega_1(\tau - 2t)] \quad \text{for} \quad \tau \in [t, t+\tau_9], \quad (3.39)$$

$$\cos\left[\int_0^t(-\omega_1)dt' + \int_{-\tau_9}^{0}(-\omega_1)dt' + \int_{t-\tau}^{-\tau_9}\omega_1 dt'\right] + \cos\left[\int_0^t(-\omega_1)dt' + \int_{-\tau_9}^{0}(\omega_1)dt' + \int_{t-\tau}^{-\tau_9}(-\omega_1)dt'\right] = \cos[\omega_1(\tau - 2t - 2\tau_9)] + \cos[\omega_1(\tau - 2\tau_9)] \quad \text{for} \quad \tau \in [t+\tau_9, t+\tau_9+\tau_8], \quad (3.40)$$

...

$$\cdots = 2\cos[\omega_1(\tau - 8\tau_R)] \quad \text{for} \quad \tau \in [t + 7\tau_R + \tau_9 + \tau_8 + \cdots + \tau_2, 8\tau_R]. \quad (3.41)$$

Equations 3.37–3.41 are valid during the first pulse $0 \le t \le \tau_1$. When $\tau_1 \le t \le \tau_2$, i.e., within the second pulse of the HDR sequence, an analogous sum of 73 terms can be written as follows:

$$j_0^{(2)}(t) = \int_0^{t-\tau_1} G(t,\tau)d\tau + \int_{t-\tau_1}^{t} G(t,\tau)d\tau + \int_t^{t+\tau_9} G(t,\tau)d\tau$$
$$+ \int_{t+\tau_9}^{t+\tau_9+\tau_8} G(t,\tau)d\tau + \cdots + \int_{t+7\tau_R+\tau_9+\tau_8+\cdots+\tau_3}^{8\tau_R} G(t,\tau)d\tau. \quad (3.42)$$

In other terms, referring to Fig. 3.7b, one has to shift the integration interval $[0, 8\tau_R]$ so that the points $\tau = 0$ and $\tau = 8\tau_R$ fall in the second pulse of two consecutive HDR blocks. In each integration interval of Eq. 3.42 the phases $\varphi_k(t,\tau)$ $(k = I, S)$ are constant and can be calculated in analogy to Eqs. 3.38–3.41. The procedure outlined above is iterated to evaluate $j_0^{(3)}(t), \ldots, j_0^{(72)}(t)$.

It is now useful to introduce the notation δ_i, with $i = 1, \ldots, 72$, to denote the time interval of the ith pulse of the HDR WALTZ-32 sequence, i.e., $\delta_1 = [0, \tau_1]$, $\delta_2 = [\tau_1, \tau_1 + \tau_2]$,...,$\delta_{72} = [7\tau_R + \tau_1 + \tau_2 + \cdots + \tau_7 + \tau_8, 8\tau_R]$. The effective multiple-quantum cross-relaxation rate can be obtained by integrating Eq. 3.36 over the whole HDR WALTZ-32 block,

$$\mu_{MQ}^{WALTZ} = \frac{1}{8\tau_R}\int_0^{8\tau_R} \mu_{MQ}^{WALTZ}(t)dt = \frac{1}{2}\overline{\Delta\omega}^2 \frac{1}{1 - e^{-8\tau_R/\tau_{ex}}} \frac{1}{8\tau_R} \sum_{i=1}^{72} \int_{\delta_i} j_0^{(i)}(t)dt, \quad (3.43)$$

in which the integral in the rightmost term of the above equation is taken over the 72 time intervals δ_i. In practice, each term $j_0^{(i)}(t)$ $(i = 1, \ldots, 72)$ is given by a sum of 73 integrals, thus one has to calculate $72 \times 73 = 5256$ integrals. Explicit calculations were carried out with Mathematica 7 on a MacBook Pro equipped with a 2.66 GHz Intel Core 2 Duo processor and 4 GB of RAM. Our serial implementation of the integration procedure led in about 2 h to the evaluation of the functions $j_0^{(i)}(t)$ and the integrals in Eq. 3.43. We obtained,

$$\mu_{MQ}^{WALTZ} = \frac{\overline{\Delta\omega}^2 \tau_{ex}}{1 + \omega_1^2\tau_{ex}^2}\left[y(\omega_1, \tau_{ex}) + \frac{e^{-8\tau_R/\tau_{ex}}\mathbb{F}_2(\omega_1, \tau_{ex})}{\frac{8\tau_R}{\tau_{ex}}\left(1 - e^{-8\tau_R/\tau_{ex}}\right)\left(1 + \omega_1^2\tau_{ex}^2\right)}\right], \quad (3.44)$$

where $y(\omega_1, \tau_{ex}) = e^{-8\tau_R/\tau_{ex}}\mathbb{F}_1(\omega_1, \tau_{ex}) / \left[192\left(1 - e^{-8\tau_R/\tau_{ex}}\right)\right]$ and $\mathbb{F}_{1,2}$ are complicated functions of ω_1 and τ_{ex} (see Appendix A). The expression above can be simplified significantly by some approximations.

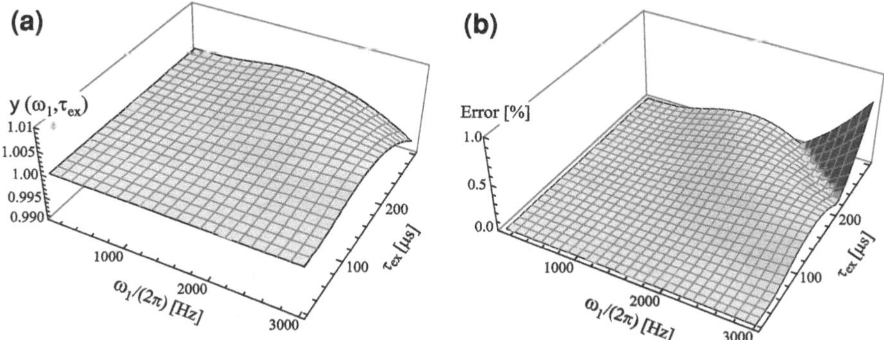

Fig. 3.8 Copyright © 2011 WILEY-VCH Verlag GmbH & Co. KGaA, Weinheim. Calculated behavior of (**a**) the function $y\,(\omega_1, \tau_{ex})$ of Eq. (3.44) and of (**b**) the error $\varepsilon = |1 - y\,(\omega_1, \tau_{ex})|\,/y\,(\omega_1, \tau_{ex})$ as a function of the *rf* amplitude ω_1 and the exchange time τ_{ex}

Indeed, one may verify by explicit numerical calculations (see Fig. 3.8) that $y\,(\omega_1, \tau_{ex}) \approx 1$. This approximation is fulfilled to a tolerance $\varepsilon < 1\,\%$ on the condition that $k_{ex} > \omega_1/(2\pi)$, which is usually well fulfilled in the fast exchange limit, with a maximum accessible *rf* amplitude $\omega_1/(2\pi) \approx 3\,\text{kHz}$. The tolerance ε drops below $0.1\,\%$ if $k_{ex} > 3\omega_1/(2\pi)$.

On the other hand, the approximation is no longer valid when $k_{ex} < \omega_1/(2\pi)$. In the limits discussed above, we may rewrite Eq. 3.44 as

$$\mu_{MQ}^{WALTZ} = \frac{\overline{\Delta\omega}^2 \tau_{ex}}{1 + \omega_1^2 \tau_{ex}^2} \left[1 + \frac{e^{-8\tau_R/\tau_{ex}} \mathbb{F}_2(\omega_1, \tau_{ex})}{\frac{8\tau_R}{\tau_{ex}} \left(1 - e^{-8\tau_R/\tau_{ex}}\right)\left(1 + \omega_1^2 \tau_{ex}^2\right)} \right]. \tag{3.45}$$

This equation has the same structure as the well-known exchange-induced rotating-frame single-quantum relaxation rate $R_{1\rho}$ under continuous-wave spin lock with constant amplitude and phase (see Sect. 3.3), modified by a correction term ($\propto\mathbb{F}_2$) that takes into account the effect of our alternated-phase double-resonance scheme. The function $\mathbb{F}_2(\omega_1, \tau_{ex})$ can also be simplified if exchange is fast compared to the *rf* amplitude ω_1. Algebraic manipulations show that $\mathbb{F}_2(\omega_1, \tau_{ex}) = A + B\omega_1\tau_{ex} + C\omega_1^2\tau_{ex}^2$, where the coefficients A, B and C are linear combinations of terms such as $\exp\left[\frac{n\pi}{2\omega_1\tau_{ex}}\right]$ and n is an integer between 0 and 192. The expressions of A, B and C are given explicitly in Appendix A. It is straightforward to verify numerically that $[A + B\omega_1\tau_{ex}]\,/\left[A + B\omega_1\tau_{ex} + C\omega_1^2\tau_{ex}^2\right] < 2\,\%$, provided $k_{ex} > \frac{1}{2}\omega_1/(2\pi)$. It is therefore possible to retain only the second-order term $C\omega_1^2\tau_{ex}^2$ in the expansion of $\mathbb{F}_2(\omega_1, \tau_{ex})$. This is shown in Fig. 3.9a.

Finally, in the expression of the coefficient C we may retain only the dominant highest-order term with $n = 192$, which turns out to be $138e^{8\tau_R/\tau_{ex}}$. This is the worst approximation in the procedure described here, leading to a relative error that can be as large as 8 %, as shown in Fig. 3.9b. With these approximations, the effective multiple-quantum cross-relaxation rate can be written as

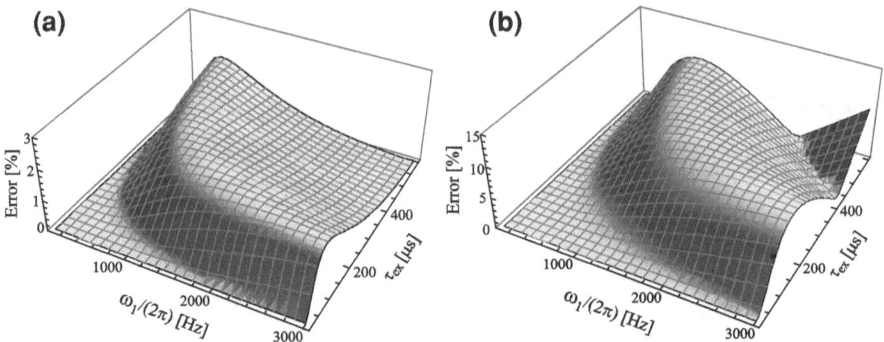

Fig. 3.9 Copyright © 2011 WILEY-VCH Verlag GmbH & Co. KGaA, Weinheim. Calculated behavior of the relative errors **a** $(A + B\omega_1\tau_{ex}) / (A + B\omega_1\tau_{ex} + C\omega_1^2\tau_{ex}^2)$ and **b** $\left|C - 138e^{8\tau_R/\tau_{ex}}\right| / C$ (see Appendix A) as a function of the *rf* amplitude ω_1 and the exchange time τ_{ex}

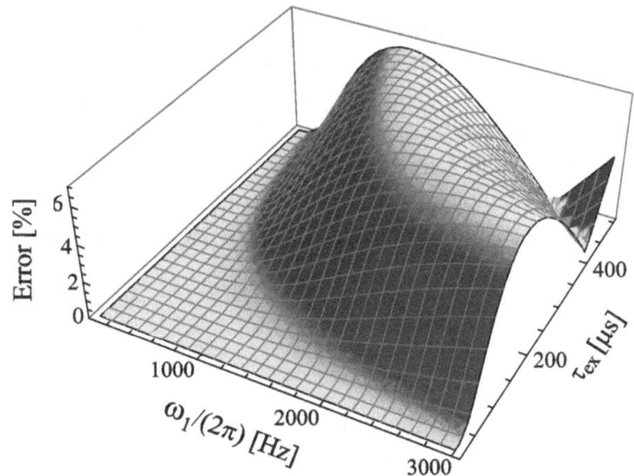

Fig. 3.10 Copyright © 2011 WILEY-VCH Verlag GmbH & Co. KGaA, Weinheim. Combined effects of all approximations leading to Eq. 3.46 compared to the function of Eq. 3.44. The relative error of Eq. 3.47 depends on the *rf* amplitude ω_1 and the exchange time τ_{ex}

$$\mu_{\text{MQ}}^{\text{WALTZ}} = \frac{\overline{\Delta\omega}^2 \tau_{ex}}{1 + \omega_1^2\tau_{ex}^2} \left[1 + \frac{23\omega_1^3\tau_{ex}^3}{16\pi \left(1 - e^{-8\tau_R/\tau_{ex}}\right)\left(1 + \omega_1^2\tau_{ex}^2\right)} \right]. \tag{3.46}$$

We may estimate the combined effects of all approximations discussed above by considering the relative error

$$\varepsilon = \frac{\left|\mu_{\text{MQ}}^{\text{WALTZ}}\,(\text{Eq.3.44}) - \mu_{\text{MQ}}^{\text{WALTZ}}\,(\text{Eq.3.46})\right|}{\mu_{\text{MQ}}^{\text{WALTZ}}\,(\text{Eq.3.44})}. \tag{3.47}$$

Its dependence on the parameters ω_1 and τ_{ex} is shown in Fig. 3.10. We can roughly estimate that the expression given in Eq. 3.46 is valid for arbitrary $k_{ex} > \omega_1/(2\pi)$ and that the error reaches a maximum $\varepsilon \approx 6\,\%$ when $k_{ex} \approx 2\omega_1/(2\pi)$, while $\varepsilon < 1\,\%$

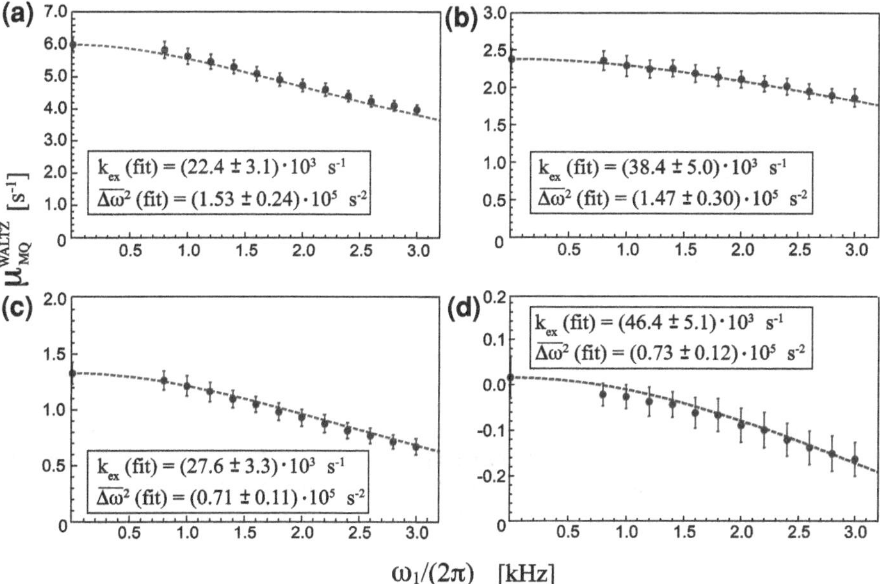

Fig. 3.11 Copyright © 2011 WILEY-VCH Verlag GmbH & Co. KGaA, Weinheim. Simulated behavior (dots) of the dispersion of the multiple-quantum cross-relaxation rate $\mu_{\mathrm{MQ}}^{\mathrm{WALTZ}}$ as a function of the *rf* field amplitude under HDR WALTZ-32 irradiation predicted for different hypotetical exchange rates, **a** $k_{ex} = 20 \cdot 10^3\ \mathrm{s}^{-1}$, **b** $k_{ex} = 40 \cdot 10^3\ \mathrm{s}^{-1}$, **c** $k_{ex} = 25 \cdot 10^3\ \mathrm{s}^{-1}$ and **d** $k_{ex} = 45 \cdot 10^3\ \mathrm{s}^{-1}$. In **a** and **b** $\overline{\Delta\omega}^2 = 1.67 \cdot 10^5\ \mathrm{s}^{-2}$, while in **c** and **d** $\overline{\Delta\omega}^2 = 0.78 \cdot 10^5\ \mathrm{s}^{-2}$. We assumed populations $p_a = 0.98$ and $p_b = 0.02$, a scalar coupling constant $J_{IS} = -90\,\mathrm{Hz}$, a correlation time $\tau_c = 6\,\mathrm{ns}$ and a static field $B_0 = 14.09\,\mathrm{T}$ (i.e., 600 MHz for ^{1}H). Moreover, we have assumed an internuclear distance $r_{IS} \approx 100\,\mathrm{pm}$, which leads to a dipolar coupling constant of about 16 kHz, and axially symmetric shielding tensors with asymmetry parameters $\Delta\sigma_I = \sigma_\parallel - \sigma_\perp = 14 \cdot 10^{-6}$ and $\Delta\sigma_S = \sigma_\parallel - \sigma_\perp = 160 \cdot 10^{-6}$, which give CSA interactions of about 8 and 9 kHz for spins I and S, respectively. These parameters are typical of a spin system with $I = {}^1$H and $S = {}^{15}$N in a protein backbone. The dashed lines are fits obtained with Equation 3.46 using k_{ex} and $\overline{\Delta\omega}^2$ as free parameters. The values of the parameters obtained from the fits are given in boxes

when $k_{ex} \geq 6\omega_1/(2\pi)$. If we take the next term ($\propto \exp\left[\frac{95\pi}{\omega_1 \tau_{ex}}\right]$) in the expansion of the coefficient C the error ε is reduced by less than 1 %. We consider this contribution to be negligible. In our experiments, where $\omega_1/(2\pi) < 3\,\mathrm{kHz}$, the uncertainty on $\mu_{\mathrm{MQ}}^{\mathrm{WALTZ}}$ due to the approximations that led to Eq. 3.46 is negligible ($\ll$1 %) when the exchange rate $k_{ex} \geq 20{\cdot}10^3\ \mathrm{s}^{-1}$. In slower regimes, with k_{ex} as small as $3{\cdot}10^3\ \mathrm{s}^{-1} \approx \omega_1/(2\pi)$, this compact analytical function has the drawback that the uncertainty on $\mu_{\mathrm{MQ}}^{\mathrm{WALTZ}}$ can reach a maximum of 6 %. Finally, another source of error is the finite asymptotic value of Eq. 3.46 when $\omega_1 \to \infty$. Indeed, one obtains $\mu_{\mathrm{MQ}}^{\mathrm{WALTZ}}(\omega_1 \to \infty) = 23\overline{\Delta\omega}^2 \tau_{ex} / \left(1536\pi^2\right)$. However, this value turns out to be only $\approx$0.1 % of the unquenched exchange contribution at $\omega_1 = 0$, i.e., $\mu_{\mathrm{MQ}}^{\mathrm{WALTZ}}(0) = \overline{\Delta\omega}^2 \tau_{ex}$, and can be thus safely neglected. In order to estimate the error introduced by the above approximations, we have also simulated numerically the dependence of the

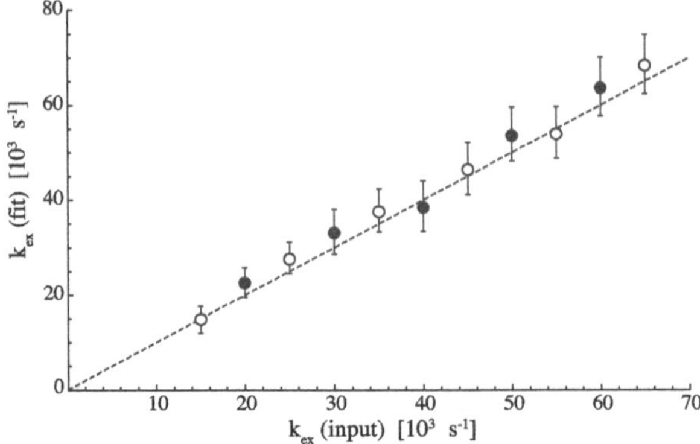

Fig. 3.12 Copyright © 2011 WILEY-VCH Verlag GmbH & Co. KGaA, Weinheim. Comparison between the values of the exchange rates k_{ex} used as input to the simulations (x-axis) and the values extracted by fitting the simulated data with Eq. 3.46 (y-axis). The *dashed line* represents the ideal case, when the two values coincide. The two sets of simulations shown here correspond to $\overline{\Delta\omega}^2 = 1.67 \cdot 10^5$ rad^{2}s^{-2} (filled circles) and $\overline{\Delta\omega}^2 = 0.78 \cdot 10^5$ rad^{2}s^{-2} (*empty circles*). All other parameters are the same as in Fig. 3.11

multiple-quantum cross-relaxation rate $\mu_{\mathrm{MQ}}^{\mathrm{WALTZ}}$ on the applied *rf* amplitude during an HDR WALTZ-32 pulse sequence under different exchange conditions. Details about the implementation of the simulations are given in Appendix B. In Fig. 3.11 we show examples of the simulated behavior of the relaxation rate of interest as a function of the applied *rf* amplitude ω_1 for different values of the exchange rate $k_{ex} = k_f/p_b = k_r/p_a$ and of the parameter $\overline{\Delta\omega}^2$. For the sake of simplicity, we show only results obtained for asymmetric populations $p_a = 0.98$ and $p_b = 0.02$. Moreover, we have assumed a scalar coupling constant $J_{IS} = -90$ Hz, typical of ^{1}H-^{15}N systems in protein backbones, a rotational correlation time $\tau_c = 6$ ns and a static field $B_0 = 14.09$ T, corresponding to a ^{1}H Larmor frequency of 600 MHz. In the simulations we assumed an uncertainty of 2 % on the exchange rate k_{ex} and on the chemical shift differences $\Delta\omega_{ab}^I$ and $\Delta\omega_{ab}^S$. These uncertainties were taken into account by a Montecarlo method and led to the error bars shown in Fig. 3.11. The simulated dispersion curves are fitted with Eq. 3.46 (dashed lines in Fig. 3.11) using a Montecarlo method with both k_{ex} and $\overline{\Delta\omega}^2$ as free parameters. The values of these parameters obtained from the fits are in excellent agreement with the input values of the simulations. This agreement is clearly shown in Fig. 3.12 for the exchange rates k_{ex}.

An important remark is that the expression given in Eq. 3.46 allows us to extract from the relaxation dispersion information about the product $p_a p_b \Delta\omega_{ab}^I \Delta\omega_{ab}^S$ of the populations and the chemical shift differences between the exchanging sites, but it does not allow one to determine these parameters separately.

References

1. Palmer AG (2004) Chem Rev 104:3623
2. Allerhand A, Thiele E (1966) J Chem Phys 45:902
3. Woessner DE (1961) J Chem Phys 35:41
4. Leigh JS (1971) J Magn Reson 4:308
5. Rance MJ (1973) J Am Chem Soc 1988:110
6. Dittmer J, Bodenhausen G (2004) J Am Chem Soc 126:1314
7. Orekov VY, Korzhnev DM, Kay LE (1886) J Am Chem Soc 2004:126
8. Korzhnev DM, Kloiber K, Kanelis V, Tugarinov V, Kay LE (2004) J Am Chem Soc 126:3964
9. Korzhnev DM, Kloiber K, Kay LE (2004) J Am Chem Soc 126:7320
10. Meiboom S (1961) J Chem Phys 34:375
11. Deverell C, Morgan RE, Strange JH (1970) Mol Phys 18:553
12. Podkorytov IS, Skrynnikov NR (2004) J Magn Reson 169:164
13. Ulzega S, Salvi N, Segawa TF, Ferrage F, Bodenhausen G (2011) Chem Phys Chem 12:333
14. Goldman M (2001) J Magn Reson 149:160
15. Kloiber K, Konrat R (2000) J Biomol NMR 18:33
16. Ulzega S, Verde M, Ferrage F, Bodenhausen G (2009) J Chem Phys 131:224503
17. Pelupessy P, Espargallas GM, Bodenhausen G (2003) J Magn Reson 161:258
18. Verde M, Ulzega S, Ferrage F, Bodenhausen G (2009) J Chem Phys 130:074506

References

Chapter 4
Experimental Methods

In the present chapter we describe in detail the experiments that are used in Chap. 5. In addition, the procedures used to analyze the experimental data are outlined.

4.1 Nitrogen-15 NMR Relaxation Experiments

Measurements of backbone ^{15}N relaxation rates have proven to be the most popular method to characterize protein dynamics in solution. The basic implementation of such experiments have been reviewed extensively [1]. Here we limit ourselves to a brief description of the pulse sequences used in the present work in addition to references to the original literature.

In the present thesis we report results obtained using the following ^{15}N NMR techniques:

- measurement of ^{15}N longitudinal auto-relaxation rates ([1, 2], Fig. 4.1);
- measurement of ^{15}N transverse auto-relaxation rates under a single echo and under a CPMG echo train ([1, 2], Fig. 4.1);
- measurement of steady-state ^{15}N-^{1}H nuclear Overhauser effects ([1, 2], Fig. 4.2);
- measurement of CSA/DD cross-correlated longitudinal [1, 3] and transverse [1, 4] cross-relaxation rates (Fig. 4.3);
- measurement of ^{15}N-^{1}H multiple-quantum differential relaxation ([5], Fig. 4.4);
- ^{15}N-^{1}H multiple-quantum CPMG relaxation dispersion ([6–9], Fig. 4.5);
- on-resonance ^{15}N $R_{1\rho}$ relaxation dispersion ([10–12], Fig. 4.6);
- off-resonance ^{15}N $R_{1\rho}$ relaxation dispersion ([10, 13], Fig. 4.6).

Our HDR pulse sequence has been presented in [14, 15] and is described in Fig. 4.7.

N. Salvi, *Dynamic Studies Through Control of Relaxation in NMR Spectroscopy*, Springer Theses, DOI: 10.1007/978-3-319-06170-2_4,

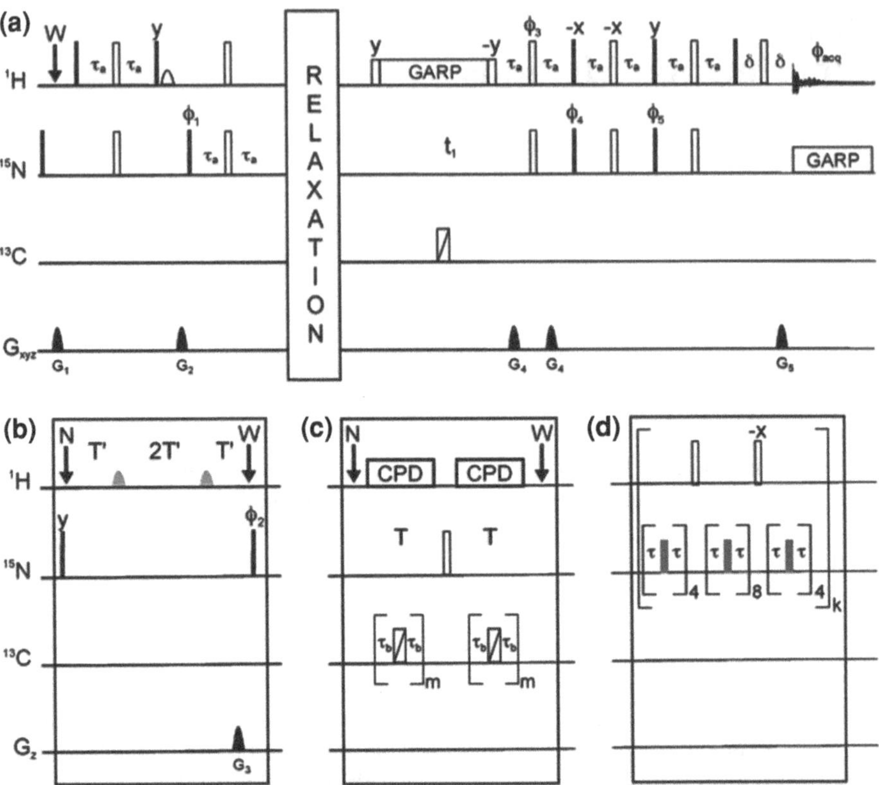

Fig. 4.1 Reproduced from [1]. Pulse sequences used for measuring ^{15}N auto-relaxation rates [1, 2]. **a** General scheme. **b** Relaxation sequence for measuring the longitudinal relaxation rate R_1, **c** relaxation sequence for measuring the transverse relaxation rate under a single echo, **d** relaxation sequence for the measurement of the transverse relaxation rate under a CPMG echo train. All *narrow* (*filled*) and (*wide*) *open rectangles* represent $\pi/2$ and π pulses, respectively. Pulse phases are along the x-axis of the rotating frame unless otherwise indicated. Composite pulse proton decoupling during the delay t_1 was performed with a GARP scheme with an *rf* amplitude of 1 kHz. The ^{13}C channel pulse can be a 500 μs smoothed CHIRP pulse [21] or omitted in samples without ^{13}C labeling, as in our experiments; the carrier can be set to 110 ppm at the center of the pulse. Composite-pulse decoupling during acquisition was performed on the ^{15}N channel with a GARP scheme [22] with an *rf* amplitude of 1 kHz. The delay τ_b is $1/(4J_{NH})$ with $4J_{NH} \approx 90$ Hz; the delay τ_b can be adjusted around 5 ms. The phase cycles were: $\phi_1 = \{$ x, −x $\}$; $\phi_3 = \{$x, x, −x, −x$\}$; $\phi_4 = \{$x, x, −x, −x $\}$; $\phi_5 = \{$−y, −y, y, y $\}$; $\phi_{\text{acq}} = \{$x, −x, −x, x $\}$. When relaxation block (**b**) is used, $\phi_2 = \{$ y, y, y, y, −y, −y, −y, −y $\}$ and $\phi_{\text{acq}} = \{$x, −x, −x, x, −x, x, x, −x $\}$. The amplitude profile of the pulsed field gradient was a sine bell shape. Their durations and peak amplitudes in the x, y, and z orientations (when triple axis gradients are available) were, respectively: G_1; 600 μs, 9.5 G/cm, 9.5 G/cm, 0; G_2; 1 ms, 0, 0, 30 G/cm; G_3; 600 μs, 15 G/cm, −15 G/cm, 0; G_4; 1 ms, 0, 0, 40 G/cm; G_5; 1 ms, 0, 0, 8.1 G/cm. Coherence selection was achieved by inverting the amplitude of the gradient G_1 and phase ϕ_1. **b** The carrier is placed at 8.2 ppm during the relaxation block; *gray bell-shaped* pulses are 1.6 ms Q3 Gaussian cascade pulses [23]. **c** WALTZ-16 decoupling should be used for ^{1}H decoupling during the relaxation block [24]. **d** *Gray rectangles* are π pulses. τ should be set to 500 μs

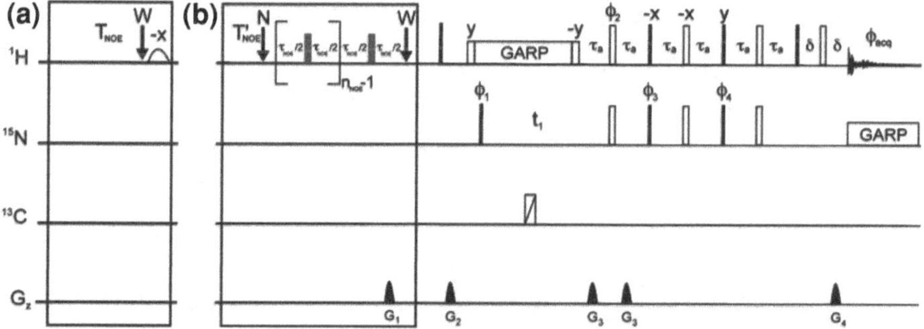

Fig. 4.2 Reproduced from [1]. Pulse sequence used for the measurement of steady-state ^{15}N-^{1}H nuclear Overhauser effects [1, 2]. For each measurement, reference (**a**) and steady-state experiments (**b**) have to be recorded in an interleaved manner. At the end of the recovery delay $T_{NOE} \geq 10$ s, the proton carrier is placed on resonance with the water signal and a very selective water-flip back pulse is applied (3 ms *sinc-shaped* or longer). In the steady-state experiments, the amide proton resonances are effectively saturated: after an optional delay $T'_{NOE} \approx 2$ s for stable detection of the lock signal, the proton carrier is placed in the center of the amide region (at 8.2 ppm), as indicated by the arrow labeled by N. The motif [delay $\tau_{NOE}/2 - \pi$ pulse-delay $\tau_{NOE}/2$] is repeated n_{NOE} times. The interpulse delay $\tau_{NOE}/2$ is typically 22 ms for globular proteins at low pH. It can be reduced to 11 ms for disordered systems at high pH, when proton exchange with the solvent is fast. The *rf* amplitude for the π pulses should be 1.5 × X kHz, where X is the ^{1}H Larmor frequency in MHz divided by 100. A gradient G_1 is applied at the end of the last $\tau_{NOE}/2$ delay to suppress all transverse components of the proton polarization. The carrier is moved on-resonance with the water signal as indicated by the arrow W. The number of cycles n_{NOE} is set so that the total duration for effective saturation is 4 s. All *narrow* (*filled*) and *wide* (*open*) *rectangles* represent $\pi/2$ and π pulses, respectively. Pulse phases are along the x-axis of the rotating frame unless otherwise stated. Proton composite pulse decoupling during the delay t_1 is performed with a GARP scheme and an *rf* amplitude of 1 kHz. Composite-pulse decoupling during acquisition is performed on the ^{15}N channel with a GARP scheme [22] and an *rf* amplitude of 1 kHz. The delay τ_a is 2.56 ms. The phase cycles are: $\phi_1 = \{y, -y\}$; $\phi_2 = \{x, x, -x, -x\}$; $\phi_3 = \{x, x, -x, -x\}$; $\phi_4 = \{-y, -y, y, y\}$; $\phi_{acq} = \{x, -x, -x, x\}$. The amplitude profile of the pulsed field gradient is a *sine bell shape*. The durations and peak amplitudes over the x, y, and z orientations are, respectively: G_1; 600 µs, 15 G/cm, 15 G/cm, 0; G_2; 1 ms, 0, 0, 25 G/cm; G_3; 1 ms, 0, 0, 40 G/cm; G_4; 1 ms, 0, 0, 8.1 G/cm. Coherence pathway selection was achieved by inverting the amplitude of the gradient G_3 and phase ϕ_4

4.2 Fitting of Dispersion Profiles to Analytical Models

During the fitting of relaxation dispersion profiles to analytical models it is often required to explore many-dimensional spaces (i.e., the models are often complex mathematical objects with many variables that must be extracted from the fit) with many local minima. Moreover, because the total exchange contribution results from a product of the chemical shifts changes and of the timescale of the process (see Eq. 3.4), the changes in these two parameters are often correlated, so that it may be challenging. from a numerical point of view, to distinguish between a slow process with a small chemical shift difference and a fast process with a large chemical shift difference.

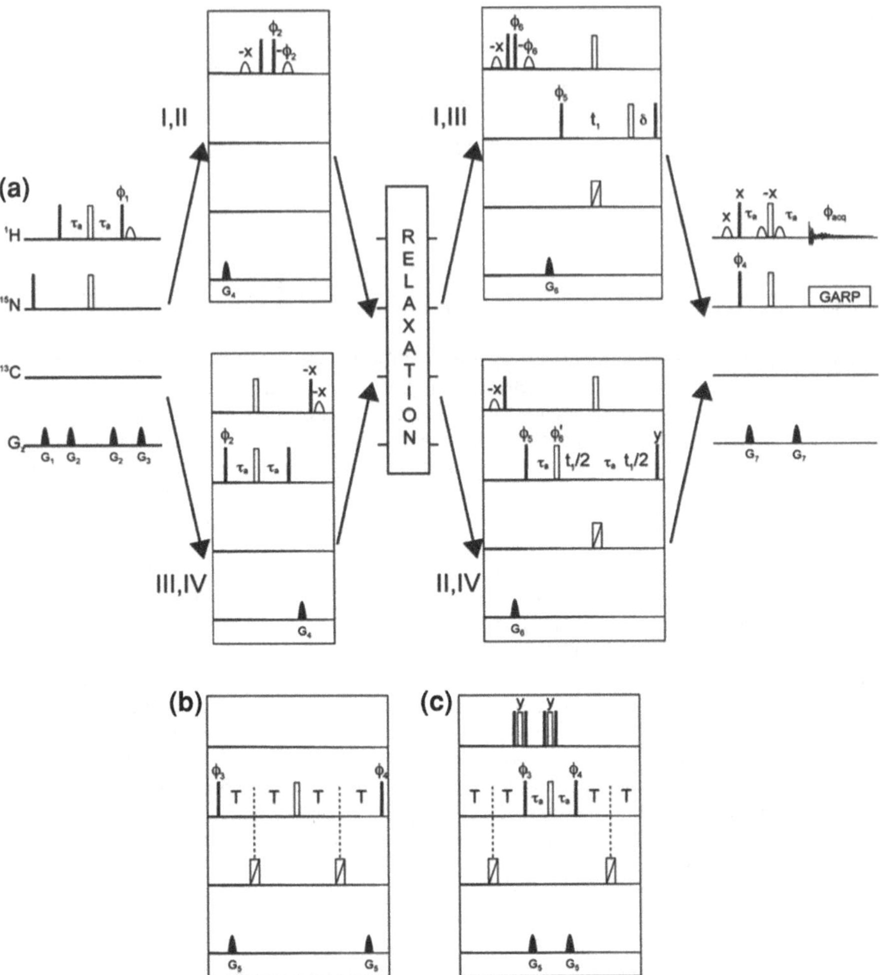

Fig. 4.3 Reproduced from [1]. Pulse sequence used for the measurement of CSA/DD cross-correlated longitudinal and transverse cross-relaxation rates [1, 3, 4]. **a** General scheme; **b** relaxation block for measuring the transverse cross-relaxation rate; **c** relaxation block for measuring the longitudinal cross-relaxation rate. Only specific elements are detailed here. The delay δ is equal to the shortest value of t_1 so that the first effective value of t_1 is zero. The phase cycles were: $\phi_1 = \{$ y, −y $\}$; $\phi_2 = \{$x, x, x, x, −x, −x, −x, −x$\}$; $\phi_3 = \{$x, x, x, x, x, x, x, x, −x, −x, −x, −x, −x, −x, −x, −x$\}$; $\phi_4 = \{$x, x, x, x, x, x, x, x, x, x, x, x, x, x, x, x, −x, −x, −x, −x, −x, −x, −x, −x, −x, −x, −x, −x, −x, −x, −x, −x$\}$; $\phi_5 = \{$x$\}$; $\phi_6 = \{$x, x, −x, −x$\}$; $\phi_6' = \{$x, x, y, y$\}$; $\phi_{\text{acq}} = \{$x, −x, −x, x, −x, x, x, −x, −x, x, x, −x, x, −x, −x, x, −x, x, x, −x, x, −x, −x, x, x, −x, −x, x, −x, x, x, −x$\}$. Gradient durations and peak amplitudes over the x, y, and z orientations were, respectively: G_1; 600 μs, 15 G/ cm, 15 G/cm, 0; G_2; 600 μs, 6.5 G/cm, 0, 0; G_3; 2 ms, 0, 0, 40 G/cm; G_4; 1 ms, −9.5 G/cm, 9.5 G/cm, 0 G/cm; G_5; 600 ms, 3.5 G/cm, 0 G/cm, 14.5 G/cm; G_6; 1 ms, −35 G/cm, −35 G/cm, −35 G/cm; G_7; 1 ms, 35 G/cm, 35 G/cm, 35 G/cm. Frequency sign discrimination was performed using States-TPPI

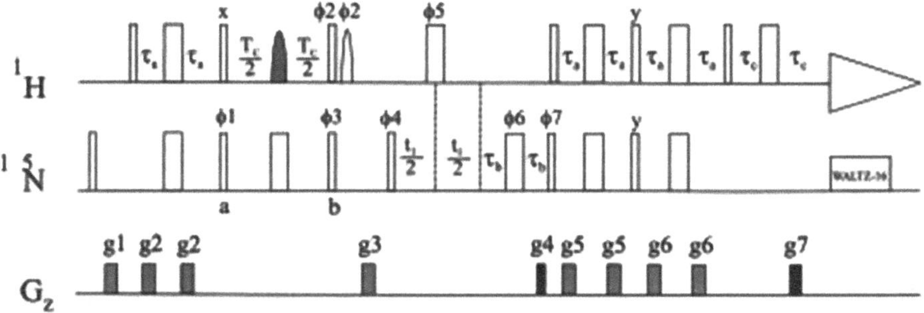

Fig. 4.4 Reproduced from [5]. Pulse sequence used for the measurement of multiple-quantum differential relaxation. *Narrow* and *wide* pulses indicate $\pi/2$ and π pulses, respectively. All pulses are applied along the x-axis, unless indicated otherwise. The ^{1}H and ^{15}N carriers are positioned on resonance with the water signal and in the center of the amide region, respectively. WALTZ decoupling [24] is achieved with a 1 kHz field. The water-selective ^{1}H $\pi/2$ pulse with phase ϕ_2 is applied as a 2.2 ms rectangular pulse, with the carrier on the water resonance. Selective inversion (*shaded pulse*) of the amide protons in the middle of the relaxation period T_c is achieved using a RE-BURP inversion pulse [25] (3.1 ms, 2.02 kHz peak *rf* amplitude). The ^{1}H carrier frequency is moved to the center of the amide region before applying the $^1H^N$ selective inversion pulse, and then it is moved back on resonance with the water signal. The values for τ_a, τ_b and τ_c are set to 2.25, 1.375 and 0.5 ms, respectively. The phase cycling was $\phi_1 = \{y, -y\}$; $\phi_2 = y$ or x; $\phi_3 = y$ or x; $\phi_4 = x$; $\phi_5 = \{x, -x\}$; $\phi_6 = \{2(x), 2(y), 2(-x), 2(-y)\}$, $\phi_7 = x$, and $\phi_{acq} = \{x, -x, -x, x\}$. Two interleaved datasets are recorded in which the phases are either set to $\phi_2 = y$; $\phi_3 = y$ (reference experiment) or $\phi_2 = x$; $\phi_3 = x$ (differential multiple-quantum relaxation experiment). Quadrature detection in the direct dimension employs the enhanced sensitivity pulsed field gradient method [26, 27] where for each value of t_1 separate data sets are recorded while alternating the signs of g_7 and ϕ_7. For each successive t_1 value, ϕ_7 and the phase of the receiver are inverted. Gradient levels are as follows: $g_1 = 1.0$ ms, 8 G/cm; $g_2 = 0.5$ ms, 8 G/cm; $g_3 = 1$ ms, 6 G/cm; $g_4 = 1.25$ ms, 30 G/cm; $g_5 = 0.15$ ms, 15 G/cm; $g_6 = 0.15$ ms, 15 G/cm; and $g_7 = 0.125$ ms, 29 G/cm

This issue can be solved by measuring relaxation dispersion profiles at different static magnetic fields: this way, the search for a value of k_{ex} that minimizes the residual sum of squares is "decoupled" from the search for the best value of $\Delta\omega$ since only the latter quantity depends on the magnetic field. However, in practice this procedure may be much too time-consuming, specially in the case of samples with limited stability.

The results of the fitting of dispersion profiles are often more reliable when a metaheuristic algorithm is employed. This class of algorithm performs a stochastic optimization of the residual sum of squares, so that the minimization steps for different parameters are effectively uncorrelated. With due precautions, the *global* minimum can be obtained. At the same time, some assumptions on the problem to be solved can be introduced, so that the convergence is faster and the numerical stability improved.

Two algorithms were implemented in Mathematica 8 [16] and/or in Matlab R2012a [17]: a genetic algorithm (GA, [18]) and a simulated annealing (SA, [19, 20]) algorithm. Appendix C contains the code that was developed. In the following descriptions, words in italics refer to variables in the code.

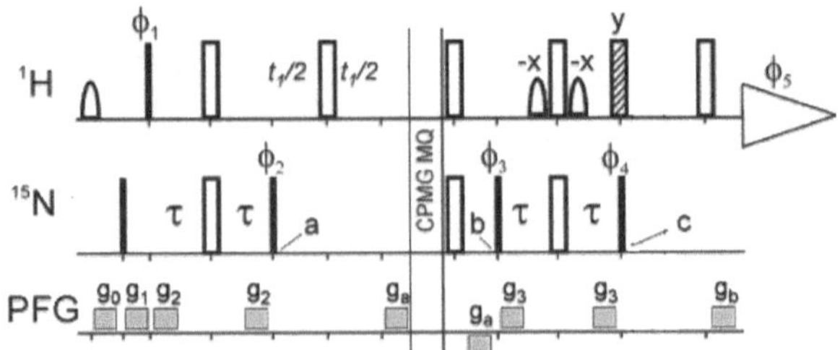

Fig. 4.5 Reprinted with permission from [8]. Copyright 2004 American Chemical Society. Pulse sequence used for ^{15}N-^{1}H multiple-quantum CPMG relaxation dispersion experiments [6–9]. All *narrow, filled* (*wide, open*) *rectangular rf* pulses are applied with flip angles of $\pi/2$ (π), respectively, along the x axis, unless indicated otherwise. The ^{1}H and ^{15}N carriers are positioned on resonance with the water signal and at center of the amide region, respectively, with the ^{1}H carrier jumped to the center of the amide region during the CPMG block (see also Fig. 3.2). At the beginning of the pulse scheme water magnetization is destroyed by the water selective EBURP-1 [25], with duration of 6 ms, followed by the gradient g_0. Further solvent suppression is achieved with the water-gate element [28] between points b and c, which makes use of rectangular water-selective $\pi/2$ pulses (1.2 ms). The delay τ is set to 2.7 ms. The duration and strengths of the field gradient pulses, which are all applied along the z axis, are g_a = 0.15 ms, 21.57 G/cm; g_b = 0.15 ms, −47.48 G/cm; g_0 = 0.3 ms, 41 G/cm; g_1 = 0.1 ms, 5.1 G/cm; g_2 = 1.9 ms, 1 G/cm; and g_3 = 0.80 ms, 4 G/cm. Gradients g_a and g_b are used to select the desired coherence transfer pathway [26, 27]. The phase cycle employed is ϕ_1 = {2(x,−x), 2(y,−y)}, ϕ_2 = {2(−y), 2(y), 2(−x), 2(x)}, ϕ_3 = { 4(x), 4(−x)}, ϕ_4 = {4(y), 4(−y)}, and ϕ_5 = {x, −x, −x, x, −x, x, x, −x}. Phase sensitive quadrature detection in the indirect dimension is achieved by recording a second free induction decay for each t_1 increment with the sign of gradients g_a inverted and the (hatched) ^{1}H π pulse immediately before point c omitted. In practice, composite pulses $90_x180_y90_x$ and $90_x180_x90_x$ are used for the first and second FIDs, respectively. The phase cycle and gradient strengths described above are used for measuring zero-quantum relaxation dispersion profiles. In the case where double-quantum profiles are obtained, ϕ_4 is inverted, and ϕ_5 = {−x, x, x, −x}. The multiple-quantum CPMG type element makes use of composite pulses of the form $90_{\psi-\pi/2}240_{\psi}90_{\psi-\pi/2}$ [29] for CPMG frequencies up to 550 Hz. Hard pulses, i.e. 180_{ψ}, are used for higher frequencies. The pulses are applied simultaneously on ^{1}H and ^{15}N *rf* channels, as shown in Fig. 3.2. The following CPMG-frequency-dependent phase scheme is employed: XY4 (0-55 Hz), ψ = {x, y, x, y}; XX (55-275 Hz), ψ = x; and XY8 for higher frequencies, ψ = {x, y, x, y, y, x, y, x}, where it is understood that for each successive π pulse in the CPMG train the phase ψ is incremented. Numerical simulations were used to check that, in each range of CPMG frequencies, the spin dynamics under the CPMG train applied using the phase cycle schemes suggested by [8] matches the evolution under a similar CPMG train of infinitely hard π pulses

The GA approach involves the evolution of a population of N_{pop} solutions termed chromosomes, where each chromosome encodes a set of fitting parameters, e.g. a value for τ_{ex}, $\overline{\Delta\omega}$ and μ_0 in the case of HDR. These values are randomly chosen inside a user-defined range. A fitness value (or cost) is assigned to each chromosome by simulating a dispersion profile and evaluating the χ^2 with respect to the experimental profile. The N_{keep} chromosomes with the best fitness values are retained, while the remaining $N_{pop} - N_{keep}$ are replaced by the next generation. For the next generation, some of the chromosomes are chosen to be the parents of the offspring, in proportion to their fitness values, using a procedure called tournament selection.

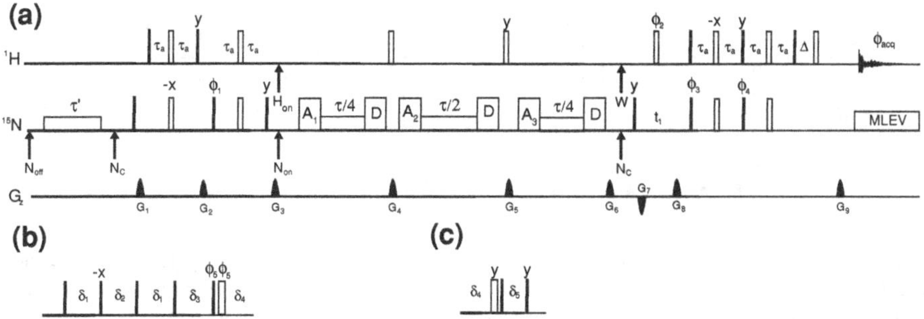

Fig. 4.6 Pulse sequence used for ^{15}N R$_{1\rho}$ relaxation dispersion experiments [10–13]. **a** General scheme. *Narrow* (*filled*) and *wide* (*open*) *rectangles* represent $\pi/2$ and π pulses, respectively. Pulse phases are along the x axis of the rotating frame unless otherwise mentioned. *Open low rectangles* indicate spin-lock periods, always applied with phase x. The first spin-lock period, before the INEPT, is used for temperature compensation. Its duration $\tau' = (\tau_{MAX}\omega^2_{1,MAX} - \tau\omega_1^2)/\omega_1^2$ is chosen so that sample heating does not depend on τ and ω_1. Arrows signal the change of carrier frequencies according to the following nomenclature: W, ^{1}H carrier in resonance with the water signal ($\approx$4.7 ppm); H$_{on}$, ^{1}H carrier in the center of the amide region (8.2 ppm); N$_{on}$, ^{15}N carrier at the frequency chosen for the spin-lock; N$_{off}$, ^{15}N carrier far-off resonance (2500 ppm); N$_c$, ^{15}N carrier in the center of the amide region (118 ppm). Delays are: $\tau_a = 1/(4J_{NH})$; Δ is set equal to the duration of G_9. The phase cycle is: ϕ_1 = {x, −x}; ϕ_2 = { 2(x), 2(−x)}; ϕ_3 = {2(x), 2(−x)}; ϕ_4 = {2(−y), 2(y)}; ϕ_{acq} = {(x, −x, −x, x, −x, x, x, −x) 2(−x, x, x, −x, x, −x, −x, x) (x, −x, −x, x, −x, x, x, −x)}. The gradients are sine-shaped of length (in ms) equal to: $G_1 = 2.0$; $G_2 = 2.0$; $G_3 = 2.0$; $G_4 = 1.0$; $G_5 = 1.0$; $G_6 = 1.0$; $G_7 = 2.0$; $G_8 = 2.0$; $G_9 = 2.0$. Their strength (in G/cm) is equal to: $G_{1z} = 14.7$; $G_{2z} = 35$; $G_{3z} = 14.7$; $G_{4z} = 14$; $G_{5z} = 17.5$; $G_{6z} = 21.7$; $G_{7z} = 56$; $G_{8z} = -G_{7z}$; $G_{9z} = 11.34$. The alignment along the direction of the effective field of the spin-lock is obtained by using the sequence in (**b**), where the delays are: $\delta_1 = 1/(2\omega_1) - 2/\omega_{HP}$, where ω_{HP} is the *rf* amplitude used for the hard pulses; $\delta_2 = \delta/\omega_1 - 2/\omega_{HP}$, in which the scaling factor δ is determined as described in [12] so as to optimize the alignment for nuclei within a specified frequency range from the carrier, in our case, between $\pm\omega_1$ from the carrier of the spin-lock field; $\delta_3 = \delta/(2\omega_1) - 2/\omega_{HP}$; $\delta_4 = 1/\omega_{HP}$. The three repetitions of the block differ for the phase cycle: in A_1, ϕ_5 = { 4(y), 4(−y)}; in A_2, ϕ_5 = { 8(y), 8(−y)}; in A_3, ϕ_5 = { 16(y), 16(−y)}. After each spin-lock period, the magnetization is aligned along the z-axis by the block D in (**c**). The delays are $\delta_4 = 1/\omega_{HP}$ as in (**b**); $\delta_5 = 1/\omega_1 - 2/\omega_{HP}$. In the case of off-resonance experiments, the alignment scheme is replaced by tanh/tan adiabatic sweeps, with a duration of 10 ms and a frequency sweep beginning at −15 kHz from the carrier frequency

The τ_{ex}, $\overline{\Delta\omega}$ and μ_0 values for each chromosome in the offspring are the average of the values of the parents. For each generation, random mutations are applied to N_{mut} components of the offspring, and their fitness values are then calculated. This process is repeated up to *MaxIter* times and the population of individuals evolves toward higher fitness solutions. The iterative process is stopped after *MaxIter* iterations or when the convergence criterion is reached (i.e. when the ratio between the standard deviation of the costs of the best N_{keep} chromosomes and the smallest cost is lower than *Prec*). The best-fit parameters are then given by the chromosome with the best fitness, and, for each parameter, the uncertainty is estimated as the standard deviation over the best N_{keep} chromosomes.

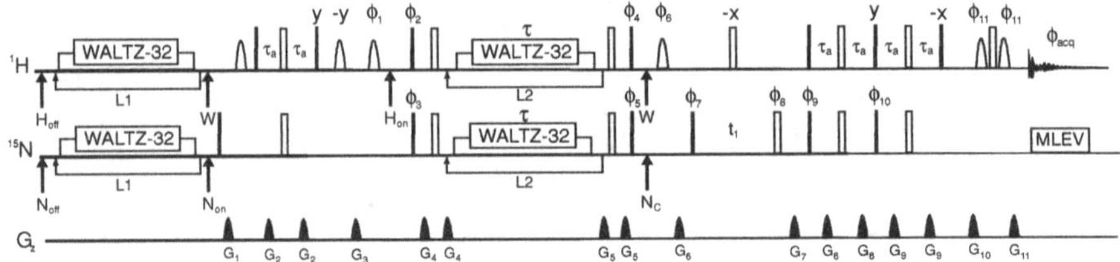

Fig. 4.7 Pulse sequence used for the measurement of multiple-quantum cross-relaxation rates under HDR WALTZ-32 irradiation [14, 15]. *Narrow (filled)* and *wide (open) rectangles* represent $\pi/2$ and π pulses, respectively. Pulse phases are along the x axis of the rotating frame unless otherwise mentioned. The *bell-shaped pulse* represents a 1 ms sinc pulse. The delay τ_a is set to $1/|4J_{NH}|$, with $J_{NH} = -90$ Hz. The recycle delay is 1.5 s. The WALTZ-32 blocks are repeated L2 times during the relaxation delay. The sequence starts with temperature compensation on both channels with WALTZ-32 blocks repeated L1 times, so that $L1 + L2 = 3$. During the temperature compensation the carrier frequencies are set far off-resonance (i.e., 320 ppm for ^{1}H and 2,500 ppm for ^{15}N), as indicated by the arrows labeled H_{off} and N_{off}. Arrows signal the change of carrier frequencies to 4.7 ppm (W) on the ^{1}H channel and on resonance with the NH^N pair under study on the ^{15}N channel (N_{on}). Before the relaxation delay, the proton carrier is placed on resonance (H_{on}). Carrier frequencies are shifted to W and N_C (118 ppm) after the relaxation period. For each spectrum, 384 transients are accumulated. The pulsed field gradients have *sine-bell-shaped* amplitude profiles. Their durations and peak amplitudes over the z axis are, respectively: $G_1 = 750$ μs, 11.5 G/cm; $G_2 = 750$μs, 4.5 G/cm; $G_3 = 750$μs, 15.5 G/cm; $G_4 = 750$μs, 13.5 G/cm; $G_5 = 750$μs, 9.5 G/cm; $G_6 = 1.4$ ms, 35 G/cm; $G_7 = 1$ ms, 40 G/cm; $G_8 = 750$μs, 14.5 G/cm; $G_9 = 750$μs, 10.5 G/cm; $G_{10} = 1$ ms, 39.05 G/cm; $G_{11} = 1$ ms, 39.05 G/cm. The phase cycles were: $\phi_1 = 8\{-y\}, 8\{y\}$; $\phi_6 = 16\{x\}, 16\{-x\}$; $\phi_7 = x$; $\phi_8 = \{x, x, y, y, -x, -x, -y, -y\}$; $\phi_9 = x$; $\phi_{10} = y$; $\phi_{acq} = 2\{x, -x, -x, x\}, 4\{-x, x, x, -x\}, 2\{x, -x, -x, x\}$. Symmetrical reconversion is used so that all four relaxation pathways between $2H_xN_x$ and $2H_yN_y$ are recorded. To generate the multiple-quantum coherence $2H_xN_x$ at the beginning of the relaxation period, we have $\phi_2 = \phi_3 = y$, while $\phi_2 = \phi_3 = x$ to generate the multiple-quantum coherence $2H_yN_y$. To detect the multiple-quantum coherence $2H_xN_x$ at the end of the relaxation period we have $\phi_4 = \phi_5 = y$, whereas $\phi_4 = \phi_5 = x$ to detect the multiple-quantum coherence $2H_yN_y$

The name and inspiration for SA come from annealing in metallurgy, a technique involving heating and slow cooling of a material to reduce defects in its structure. This notion of slow cooling is implemented in the SA algorithm as a decrease in the probability of accepting worse solutions as the space of possible solutions is explored. Accepting worse solutions is a fundamental property of metaheuristics because it allows for a more extensive search of the optimal solution. The method is an adaptation of the Metropolis-Hastings algorithm [30], a Monte Carlo method to generate samples of states of a thermodynamic system.

References

1. Ferrage F (2012) Methods Mol Biol 831:141
2. Kay LE, Torchia DA, Bax A (1989) Biochemistry 28:8972
3. Pelupessy P, Ferrage F, Bodenhausen G (2007) J Chem Phys 126:134508

4. Pelupessy P, Espargallas GM, Bodenhausen G (2003) J Magn Reson 161:258
5. Kloiber K, Konrat R (2000) J Biomol NMR 18:33
6. Mittermaier A, Orekhov VY, Kay LE (2005) J Am Chem Soc 127:15602
7. Korzhnev DM, Kloiber K, Kay LE (2004) J Am Chem Soc 126:7320
8. Orekhov VY, Korzhnev DM, Kay LE (1886) J Am Chem Soc 2004:126
9. Hansen DF, Vallurupalli P, Kay LE (2008) J Phys Chem B 112:5898
10. Massi F, Johnson E, Wang CY, Rance M, Palmer AG (2004) J Am Chem Soc 126:2247
11. Paquin R, Ferrage F, Mulder FAA, Akke M, Bodenhausen G (2008) J Am Chem Soc 130:15805
12. Hansen DF, Kay LE (2007) J Biomol NMR 37:245
13. Mulder FAA, de Graaf RA, Kaptein R, Boelens R (1998) J Magn Reson 131:351
14. Salvi N, Ulzega S, Ferrage F, Bodenhausen G (2012) J Am Chem Soc 134:2481
15. Verde M, Ulzega S, Ferrage F, Bodenhausen G (2009) J Chem Phys 130:074506
16. Wolfram Research Inc (2010) Mathematica Edition: Version 8.0; Wolfram Research Inc
17. MATLAB, version 7.14.0 (R2012a); The MathWorks Inc, 2012
18. Melanie M (1999) An introduction to genetic algorithms. MIT Press, Cambridge
19. Kirkpatrick S, Gelatt CD, Vecchi MP (1983) Science 220:671
20. Černý V (1985) J Optim Theory Appl 45:41
21. Böhlen J-M, Bodenhausen G (1993) J Magn Resona A 102:293
22. Shaka AJ, Barker PB, Freeman R (1985) J Magn Reson 64:547
23. Emsley L, Bodenhausen G (1990) Chem Phys Lett 165:469
24. Shaka AJ, Keeler J, Frenkiel T, Freeman R (1983) J Magn Reson 52:335
25. Geen H, Freeman R (1991) J Magn Reson 93:93
26. Kay LE, Keifer P, Saarinen T (1992) J Am Chem Soc 114:10663
27. Schleucher J, Sattler M, Griesinger C (1993) Ang Chem Int Ed Eng 32:1489
28. Piotto M, Saudek V, Sklenár V (1992) J Biol NMR 2:661
29. Freeman R, Kempsell SP, Levitt MH (1980) J Magn Reson 38:453
30. Metropolis N, Rosenbluth AW, Rosenbluth MN, Teller AH, Teller E (1087) J Chem Phys 1953:21

Chapter 5
Experimental Results

5.1 Internal Dynamics in Human Ubiquitin

Internal motions in ubiquitin have been studied extensively by NMR. Indeed, ubiquitin has been used for two decades as a standard for biomolecular NMR, and new methods have been frequently validated with experiments on ubiquitin. Motions faster than overall tumbling have been characterized in detail [1, 2] and can now be reproduced with good accuracy by MD simulations [3, 4]. Motions on microsecond-millisecond time scales have been identified [5–8], and their time scales have been determined for a few residues using several techniques over a range of pH and temperatures [5, 9–12]. Motions on time scales slower than the overall tumbling but faster than a few μs have been explored more recently, in particular by exploiting residual dipolar couplings [13–15] in combination with computational approaches [16, 17]. A normal mode has been identified as the main source of conformational diversity in ubiquitin, whether it is free or bound in complexes, suggesting a conformational selection mechanism for binding [16].

At least two interaction sites have been identified in ubiquitin. For instance, the binding of ubiquitin to the exchange factor for Rab5 Rabex-5 occurs at two distinct sites: the inverted ubiquitin-interacting motif, which binds the Ile 44 hydrofobic patch on ubiquitin (in red in Fig. 5.1), which comprises residues L8, R42, I44, A46, G47, K48, K63, H68 and V70; and the N-terminal zinc finger, which interacts with polar residues (in green in Fig. 5.1) such as S20, G53, R54, D58, Y59 and N60 [18–21].

In this section we use the heteronuclear double resonance method to determine the time scales of microsecond motions in ubiquitin using a large number of probes. We show that most motions occur on the same time scale. We suggest that a small tilt of the α-helix, which may be modulated by the dynamics of H-bonds at both ends, may explain our observations.

All the experiments reported here were carried out on a sample of perdeuterated and uniformly ^{15}N labeled human ubiquitin (1.5 mM, pH 6.8) at 280 K (unless otherwise specified) on an 18.79 T (800 MHz for ^{1}H) Bruker Avance spectrometer equipped with a TXI cryoprobe with z-axis gradients.

N. Salvi, *Dynamic Studies Through Control of Relaxation in NMR Spectroscopy*, Springer Theses, DOI: 10.1007/978-3-319-06170-2_5,

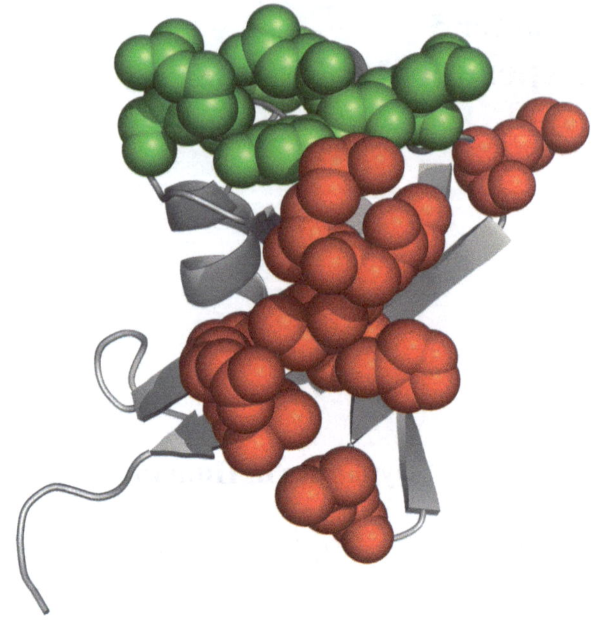

Fig. 5.1 Binding surfaces in human ubiquitin. A solution-state structure of ubiquitin [22] (pdb code 1d3z), and PYMOL [23] were employed to generate the image

5.1.1 Identifying Chemical Exchange from the Relaxation of Single-Quantum Coherences

In Fig. 5.2a, the contributions of chemical exchange to the transverse relaxation of backbone ^{15}N nuclei (R_{ex}) are derived from the comparison of transverse auto-relaxation rates (R_2) and cross-correlated cross-relaxation rates (η_{xy}) [24]. The rates R_2 were measured using a Carr-Purcell-Meiboom-Gill (CPMG) echo train, and the rates η_{xy} were determined using the symmetrical reconversion method [25]. The expression $R_{ex} = R_2 - \kappa\eta_{xy}$ was used with $\kappa = 1.31$. The value of κ was derived from the correlation of R_2 and η_{xy} excluding the residues with the 15 lowest and the 15 highest values of R_2. The black dashed line in Fig. 5.2a marks the average value of R_{ex}, i.e. values above this threshold have to be considered outliers with significant contributions of μs-ms exchange processes. Chemical exchange is therefore observed in the hydrophobic patch (Leu8, Thr9, Val70) at the N-terminus of helix $\alpha1$ (Ile23 and Asn25; the signal of Glu24 being too weak) and in the $\beta4$-$\alpha2$ loop (Thr55). Chemical exchange also seems to be detected for the very mobile residues Leu73 and Arg74; these contributions are likely due to proton exchange.

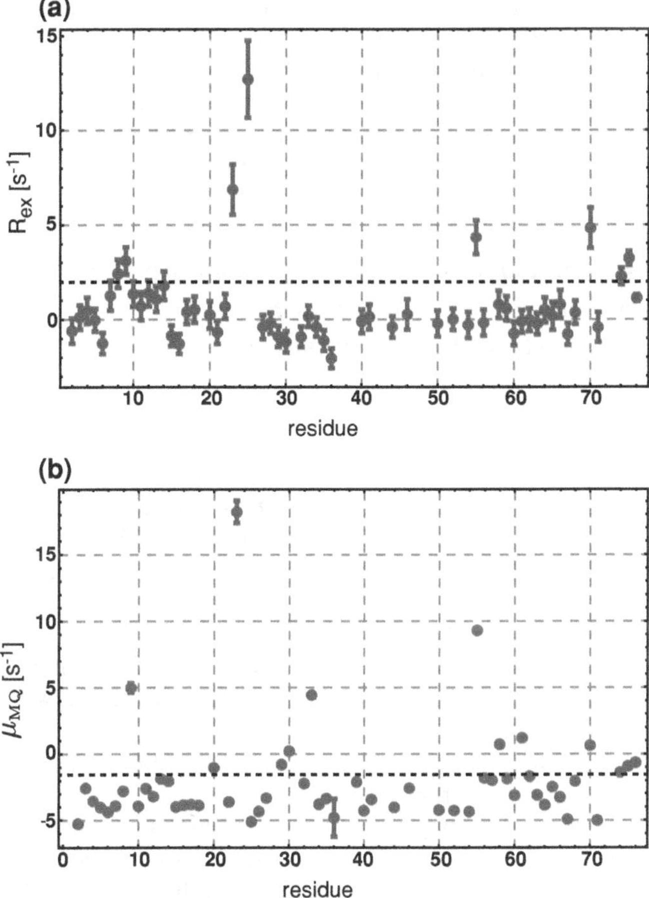

Fig. 5.2 Reprinted with permission from [26]. Copyright 2012 American Chemical Society. Identification of chemical exchange from relaxation rates in ubiquitin. **a** Contributions R_{ex} of chemical exchange to the transverse nitrogen-15 relaxation rates at 280 K and 18.79 T (800 MHz for ^{1}H). **b** Multiple-quantum cross-relaxation rates measured with the pulse sequence of Kloiber and Konrat [27] at 280 K and 18.79 T (800 MHz for ^{1}H)

5.1.2 Identifying Chemical Exchange from the Relaxation of Multiple-Quantum Coherences

Interestingly, the signature of chemical exchange is significantly more pronounced for the cross-relaxation rates of multiple-quantum coherences than for the auto-relaxation rates of ^{15}N single-quantum coherences. Indeed values of μ_{MQ} that differ significantly from the average for a given protein and isotope labeling scheme are typically interpreted by invoking a correlated chemical exchange process [5, 27].

This is the case for the values reported in Fig. 5.2b, where the average is marked by a black dashed line. In practice, all residues for which $\mu_{MQ} > 0$ can be considered to show the signature of chemical exchange. More precisely, the results of multiple-quantum spectroscopy confirm the presence of exchange for Thr9, Ile23, Thr55, and Val70 and provide further evidence of chemical exchange at the C-terminus of helix $\alpha 1$ (Ile30 and Lys33) as well as in helix $\alpha 2$ (Asp58) and in loop $\alpha 2$-$\beta 5$ (Ile61).

5.1.3 Identifying Chemical Exchange from the Temperature Dependence of the Relaxation Rates of Multiple-Quantum Coherences

In addition, the temperature dependence of μ_{MQ} offers a way to identify the presence of exchange for residues that have a small contribution to the relaxation rate of interest. Indeed, as the temperature decreases, the viscosity of water increases, which leads to an increase in the magnitude of the exchange-free cross-relaxation rates of the multiple-quantum coherences, in agreement with the general trend (see Fig. 5.3a). Contributions of fast chemical exchange to relaxation also increase with decreasing temperature. Depending on the relative signs of the chemical shift changes of nitrogen-15 and protons, these two effects will lead to an enhanced or reduced temperature dependence of μ_{MQ}. In the light of these effects, small contributions of chemical exchange can also be identified for residues Ile13, Thr14, Ser20, Lys29, Ala46, Ile56, Tyr59, Gln62, and His68 (see Fig. 5.3b).

Finally, the amide pairs of Leu43 and Phe45 show signatures of chemical exchange at 295 K, but their signals overlap at 280 K.

Figure 5.4 presents a summary of the residues in ubiquitin for which contributions of chemical exchange to single- or multiple-quantum relaxation rates have been identified. The largest effects are seen at both ends of helix $\alpha 1$ as well as at the two main interaction sites of Fig. 5.1, i.e. the hydrophobic patch and the loops and helix between $\beta 4$ and $\beta 5$.

5.1.4 Control of Temperature Variations During HDR Irradiation

Since the HDR irradiation during the relaxation delay may cause heating, we monitored the temperature changes during our experiments as follows. A series of heteronuclear single quantum coherence (HSQC) spectra were recorded between 280 and 303 K (Fig. 5.5). Differences in chemical shifts of some selected pairs of resonances were fitted to a linear function of temperature (Fig. 5.6a). These differences were then monitored throughout the measurement of a full dispersion profile at the nominal temperature of 280 K (Fig. 5.6b).

The lowest temperature was found in K&K experiments [27], i.e. at $\omega_1/(2\pi) = 0$, (279.4 ± 0.7 K), while the highest (280.0 ± 0.7 K) was measured in HDR

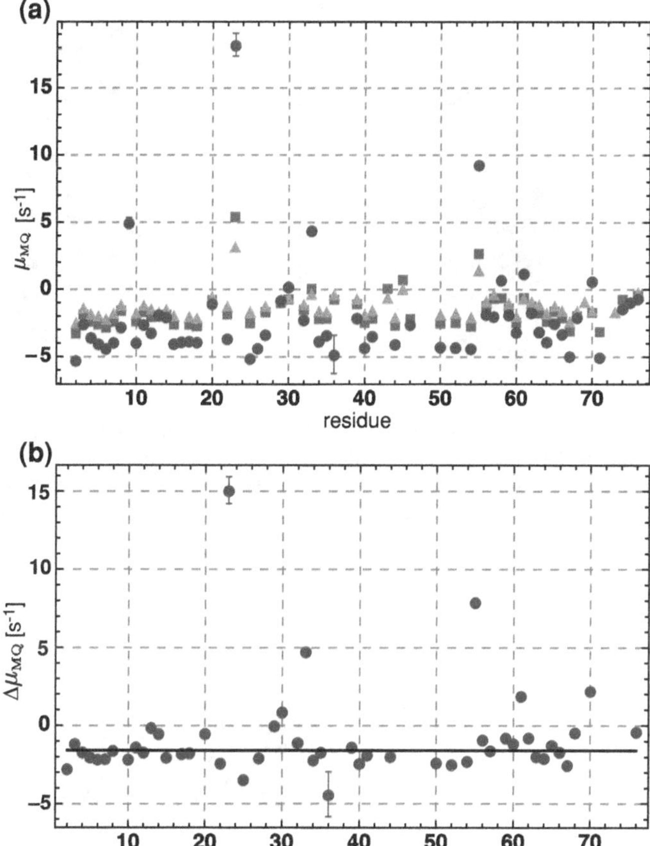

Fig. 5.3 Reprinted with permission from [26]. Copyright 2012 American Chemical Society. Temperature dependence of the relaxation rates of multiple-quantum coherences. **a** Multiple-quantum cross-relaxation rates measured with the pulse sequence of Kloiber and Konrat [27] at 280 K (*blue circles*), 295 K (*red squares*), and 303 K (*green triangles*) and 18.79 T (800 MHz for ^{1}H). **b** Difference between the multiple-quantum cross-relaxation rates measured at 280 and 303 K

experiments with the highest *rf* amplitude $\omega_1/(2\pi) = 2.9$ kHz and the longest relaxation delay $T_{rel} = 50$ ms. It is possible to conclude that such small variations do not preclude a quantitative analysis.

5.1.5 Quantifying Chemical Exchange by HDR Relaxation Dispersion

The HDR dispersion profiles, measured by using the pulse sequences of the Chap. 4, are shown in Fig. 5.7. With the exception of Thr9, all profiles show a small decrease

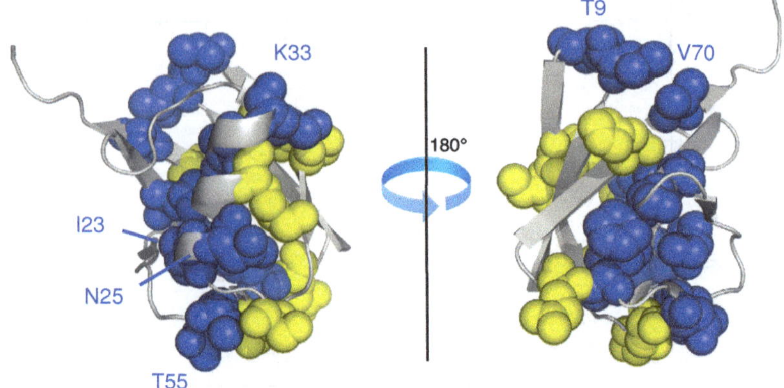

Fig. 5.4 Chemical exchange contributions to relaxation rates mapped onto the structure of ubiquitin. A solution-state structure of ubiquitin [22] (pdb code 1d3z), and PYMOL [23] were employed to generate these images. Residues rendered in *blue* show contributions to single-quantum and/or multiple-quantum relaxation rates, while residues in *yellow* feature a nontrivial temperature dependence of μ_{MQ}

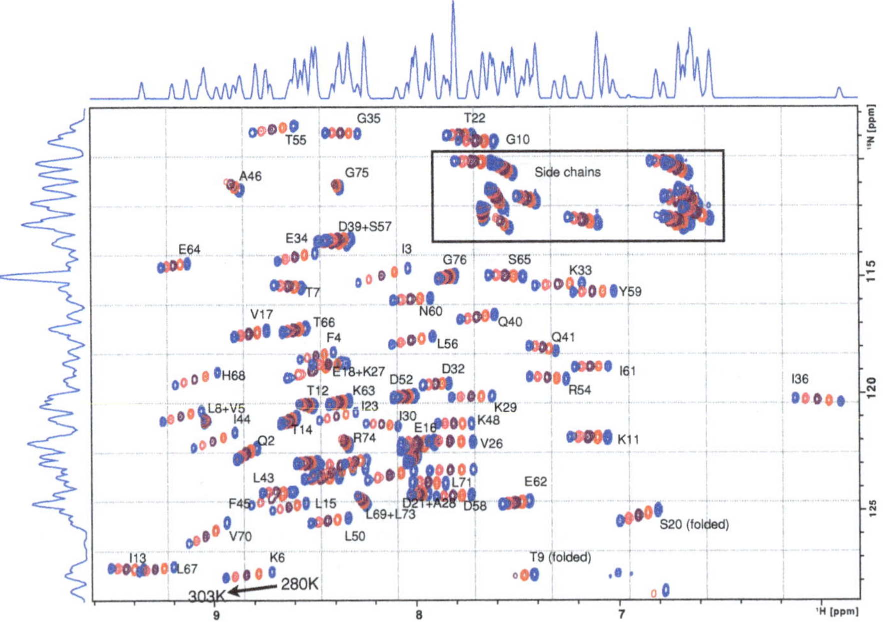

Fig. 5.5 ^{1}H, ^{15}N-HSQC spectra of human ubiquitin as a function of the temperature

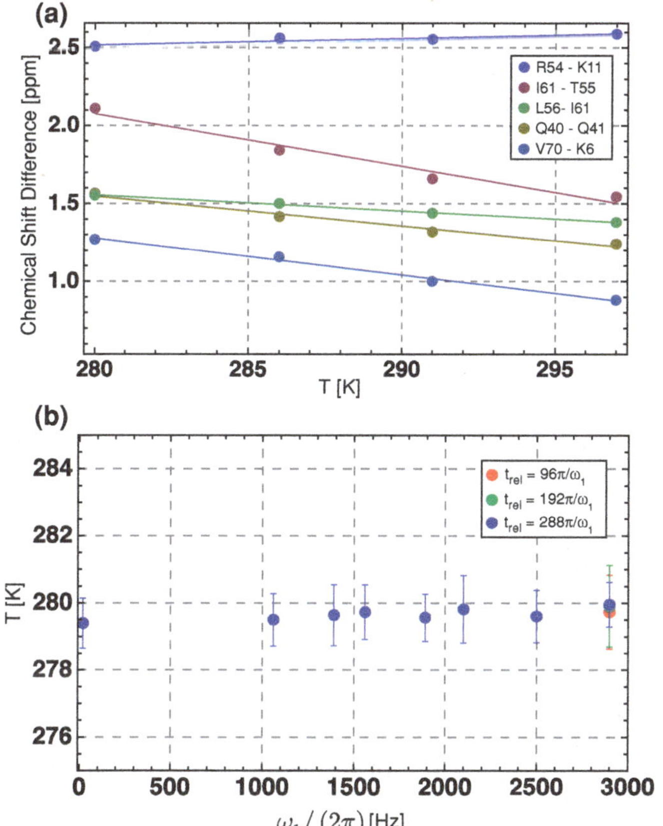

Fig. 5.6 Reprinted with permission from [26]. Copyright 2012 American Chemical Society. Control of the variations of the temperature during HDR irradiation. **a** The temperature was monitored by measuring the chemical shift difference $\Delta\delta = \sqrt{\Delta\delta_N^2 + \Delta\delta_H^2}$, where $\Delta\delta_N$ and $\Delta\delta_H$ are chemical shift differences between backbone nitrogen and amide protons of a suitable pair of residues. This parameter was measured for several pairs of residues (R54 and K11, I61 and T55, L56 and I61, Q40 and Q41, V70 and K6) in a series of HSQC spectra in the range 280–303 K. A calibration curve for each pair of residues was obtained by fitting the temperature dependence to a linear function of T. **b** Temperature variations after a relaxation delay of $96\pi/\omega_1$ (corresponding to a single HDR-WALTZ-32 block, *red*), $192\pi/\omega_1$ (two HDR-WALTZ-32 blocks, *green*) and $288\pi/\omega_1$ (three HDR-WALTZ-32 blocks, *blue*), which is the longest HDR sequence used in our experiments

of μ_{MQ} with increasing *rf* amplitude, indicating the presence of exchange processes with rates comparable to the maximum *rf* amplitudes employed.

Thus the dispersion profiles were fitted to Eq. 3.46 using the genetic algorithm described in Chap. 4 and Appendix C. The rates extracted from the individual fits of the profiles are very similar (see Table 5.1), with an average 18,800 s^{-1} and a standard deviation 900 s^{-1}.

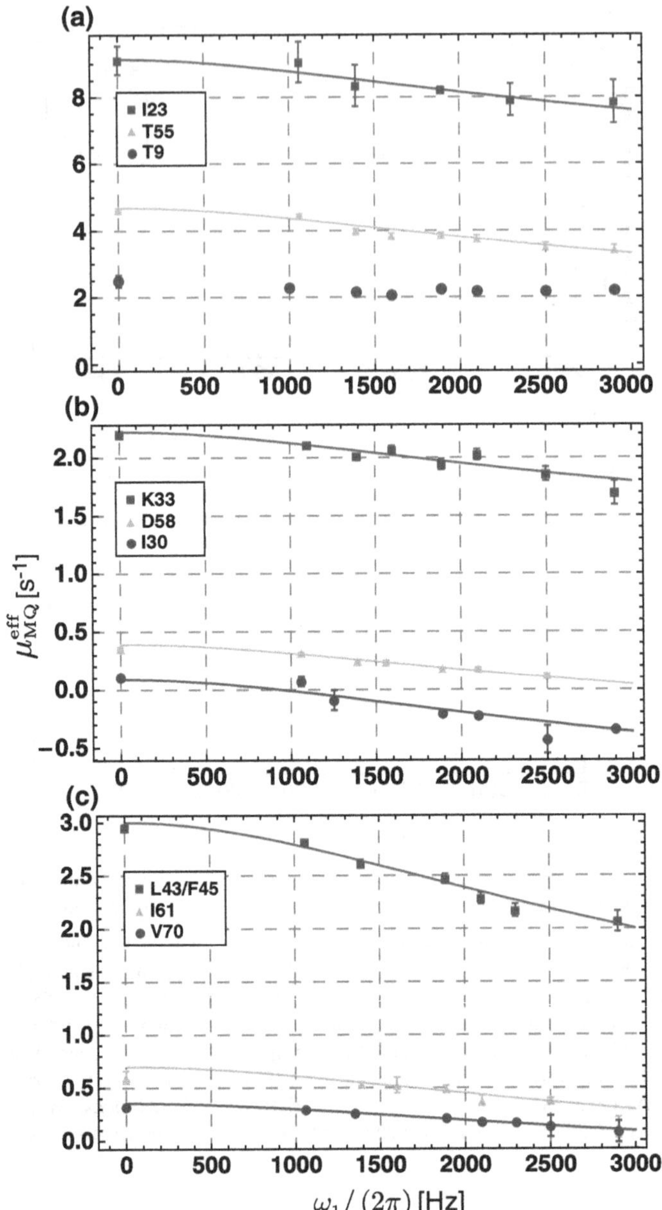

Fig. 5.7 Reprinted with permission from [26]. Copyright 2012 American Chemical Society. HDR relaxation dispersion profiles for nine amide pairs in human ubiquitin at 280 K. **a** Profiles for the NH^N pairs of Thr9, Ile23, and Thr55. **b** profiles for the NH^N pairs of Ile30, Lys33, and Asp58; **c** profiles for the NHN pairs of Ile61, Val70 and the overlapping signals of Leu43 and Phe45. The solid lines correspond to a global fit to Eq. 3.46. In the case of Thr9, μ_{MQ} does not decrease with increasing *rf* amplitude, indicating that exchange occurs on time scales faster than can be probed by our HDR experiments. Thus, Thr9 was not included in the global fitting

Table 5.1 Individual fitting parameters of HDR dispersion profiles in ubiquitin

Residue	k_{ex} [10^3 s^{-1}]	$\overline{\Delta\omega^2}$ [10^3 s^{-2}]	μ^0_{MQ} [s^{-1}]
I23	18 ± 7	65 ± 20	10 ± 4
I30	19.3 ± 1.5	23 ± 3	-2.2 ± 0.4
K33	18.5 ± 1.5	20 ± 3	2.2 ± 0.2
L43/F45	18 ± 3	50 ± 10	0.8 ± 0.4
T55	18 ± 4	60 ± 20	2.6 ± 1
D58	18.4 ± 0.5	16 ± 1	$-.0 \pm 0.2$
I61	20 ± 3	22 ± 3	-0.8 ± 0.6
V70	20.2 ± 0.6	15 ± 5	-0.8 ± 0.4

Table 5.2 Global fitting parameters of HDR dispersion profiles in ubiquitin

Residue	$\overline{\Delta\omega^2}$ [10^3 s^{-2}]	μ^0_{MQ} [s^{-1}]
I23	76 ± 9	5.1 ± 0.3
I30	23 ± 3	$-.1 \pm 0.1$
K33	20 ± 3	1.15 ± 0.08
L43/F45	49 ± 6	0.4 ± 0.2
T55	70 ± 9	1 ± 0.3
D58	18 ± 2	-0.52 ± 0.06
I61	18 ± 2	-0.26 ± 0.06
V70	12 ± 2	0.3 ± 0.04

Thus a global fitting was performed under the hypothesis that our data are compatible with dynamics that occur on a single timescale. A rate $k_{ex} = (19 \pm 1)10^3$ s^{-1} was obtained, while the other extracted parameters are reported in Table 5.2. Based on the corrected Aikake Information Criterion (AICc), the global model has a probability of >99 % of reproducing the data better than the individual fits. This time scale is in good agreement with results obtained using single-quantum relaxation dispersion [5, 12].

Identical time scales are not sufficient to prove correlated motions, as has been shown in studies of RNase A [28, 29]. Yet evidence of a common time scale invites us to speculate about possible mechanisms of correlated dynamics.

As illustrated in Fig. 5.8, the motions of the N-terminus of helix $\alpha 1$ and the $\beta 4$-$\alpha 2$ loop are coupled through H-bonds between the side chain carboxyl of Glu24 and the backbone amide proton of Gly53 [30, 31] as well as between the backbone amide proton of Ile23 and the backbone CO of Arg54 [5]. This motion is likely to be coupled with the dynamics of Thr55 and Asp58 by an H-bond between the side chain carboxyl group of Asp58 and the backbone H^N of Thr55. At the C-terminal end of helix $\alpha 1$, the H-bond between the amide proton of Lys33 and the CO of Lys29 couples the motions of the peptide planes comprising the Lys33 and Ile30 amide pairs.

Possible couplings of the motions of the N- and C- termini of helix $\alpha 1$ should also be considered. For instance, the G53A mutation leads to a significant enhancement of the contribution of chemical exchange to the transverse single-quantum relaxation of the backbone nitrogen-15 nuclei of Ala28 and Lys33 [30]. In addition, although intermolecular contacts may also be considered, the presence of a highly populated

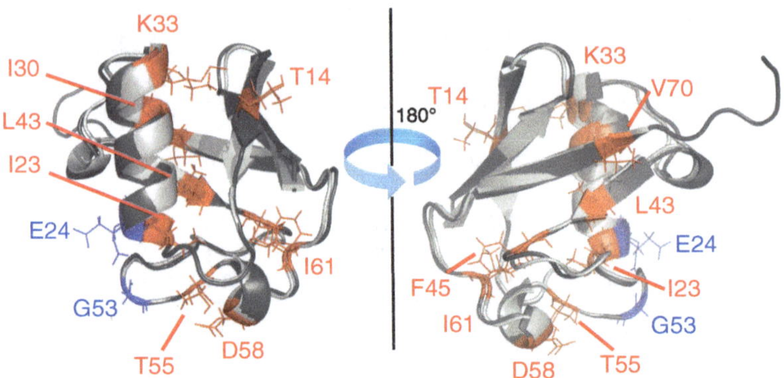

Fig. 5.8 Possible coupling of motions in human ubiquitin. The solution state (*dark gray*, pdb code 1d3z) and crystal structures (*light gray*, pdb code 3ons) of ubiquitin were aligned to minimize deviations of the β sheets

Glu24-Gly53 H-bond in the microcrystalline form of ubiquitin [31] is correlated with significant changes in the chemical shift of the backbone ^{15}N of Lys33 [32]. Comparison of the solution state (pdb code 1d3z, [22]) and microcrystalline (pdb code 3ons, [31]) structures also reveals the presence of an additional H-bond between the side-chain ammonium group of Lys33 and the backbone CO of Thr14. The existence of this H-bond has been confirmed by MD simulations and experiments in solution [33, 34]. Alignment of the β-sheets of the two structures (Fig. 5.8) offers interesting insight into the conformational transition. The helix $\alpha 1$ is slightly tilted, as if it were pulled on one side by the Glu24-Gly53 H-bond and on the other side by the Lys33-Thr14 H-bond. The distances between the Cα in the two structures are 0.5 Å for Glu24 and 1.1 Å for Lys33. The dynamic nature of helix $\alpha 1$ has been observed in several studies [35–37]. Note that the tilt between the two structures is small, on the order of 5 degrees, which would be less than previously reported [37], and could possibly be a projection of a more complex motion [16, 17].

Further correlations of motions, in particular between the β-sheet and helix $\alpha 1$, remain hypothetical. However, in contradiction to what would be expected if motions at distinct binding sites were coupled [38, 39], a study of a ternary complex of ubiquitin failed to detect any allosteric effects between the hydrophobic patch and the $\beta 4$-$\alpha 2$ loop interfaces [40]. Surprisingly, the picture of motions in ubiquitin derived from our chemical exchange study seems different from the one drawn from an analysis of RDCs [16, 17]. This underlines the diversity of motions in ubiquitin as RDCs are more sensitive to fluctuations of the major conformer, while our work focuses on transitions to a weakly populated excited state.

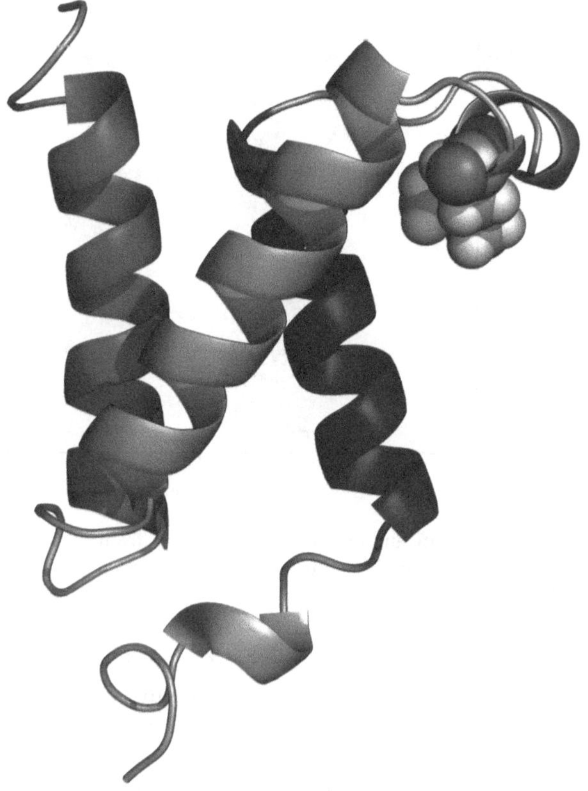

Fig. 5.9 Structure of KIX. Helices $\alpha1$, $\alpha2$ and $\alpha3$ are marked in *blue*, *green* and *red*, repectively. A sphere representation of residue L620 is provided. The pdb 1kdx [48] and PYMOL [23] were employed to generate the image

5.2 Internal Dynamics in KIX

Despite its small size (10.6 kDa), the KID-binding (KIX) domain of the CREB-binding protein (CBP) domain is made up of three α-helices in an up-down-up topology, two short 3_{10}-helices and interconnecting loops, that surround a hydrophobic core (see Fig. 5.9). In vivo, KIX is a key element in many signal pathways, interacting with a number of co-activators through two distinct binding sites [41, 42]. Interestingly, some of these co-activators, such as the kinase-inducible domain (KID) [43, 44], are (partially) unstructured proteins that undergo folding upon binding with KIX. This mechanism has an important role in the tuning of protein-protein interactions. KIX has been considered to fold through a fast appearance of an intermediate state, followed by a slower folding phase [45, 46]. The rate of the latter process is $\approx 400\,\mathrm{s}^{-1}$ at $T = 293\,\mathrm{K}$ [47].

In the present section MQ CPMG and HDR techniques are applied jointly to give a comprehensive description of all the dynamics occurring from the ms to the μs timescales in KIX. U-[^{15}N, ^{2}H] labeled samples of the KIX domain (residues 586–672) of human CBP were prepared by Sven Brüschweiler in the research groups of Prof. Robert Konrat and Dr. Martin Tollinger by bacterial growth in deuterated M9

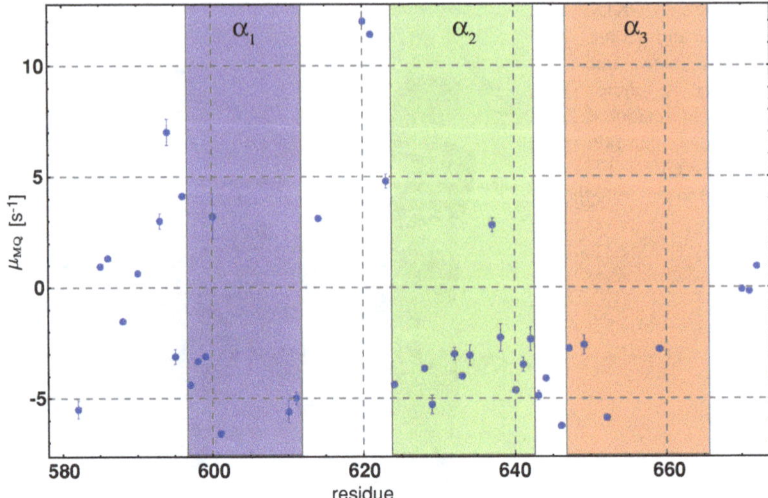

Fig. 5.10 Identification of chemical exchange from relaxation rates of multiple-quantum coherences in KIX. The multiple-quantum cross-relaxation rates were measured with the pulse sequence of Kloiber and Konrat [27]

minimal media containing $^{15}NH_4Cl$ and U-[^{12}C,^{2}H]-a-D-glucose as sole nitrogen and carbon sources, respectively, and purified as previously described [46].

All NMR samples contained 1 mM KIX, 50 mM potassium phosphate, pH 5.8, 25 mM NaCl, and 1 mM NaN_3 in 90 % D_2O / 10 % H_2O. The experiments were run in a static magnetic field $B_0 = 18.79$ T using a Bruker Avance 800 MHz spectrometer equipped with a triple resonance cryoprobe (TCI) at a temperature T = 293 K.

5.2.1 Identifying Chemical Exchange from the Relaxation of Multiple-Quantum Coherences

Evident signs of the extent of chemical exchange in KIX can be found in the cross-relaxation rates of multiple-quantum coherences of Fig. 5.10. Indeed, for many backbone amide pairs the value of μ_{MQ} is significant. In particular, while many residues in the three helices have values close to zero, significant signatures of chemical exchange are shown by residues in the loops, especially in the N-terminus (e.g., E593) and in the loop between $\alpha 1$ and $\alpha 2$ (e.g., L620 and K621).

The dynamics of these residues can be characterized by relaxation dispersion NMR. Such large chemical-exchange contributions can be due to the slow process already identified [45–47], or to the presence of additional faster processes that cannot be detected by conventional single-quantum CPMG relaxation dispersion, or to both.

5.2.2 *Quantifying Chemical Exchange by Multiple-Quantum CPMG Relaxation Dispersion*

In order to quantify contributions of the slow process to the cross-relaxation of multiple-quantum coherences, we performed MQ CPMG experiments using the method described by Kay and co-workers [49, 50]. CPMG pulse repetition rates (ν_{CPMG}) between 40 and 650 Hz were used. Exact numerical solutions of the Bloch-McConnell equations described in Sect. 3.2.2 were fitted to the experimental DQ and ZQ relaxation rates R_{DQ} and R_{ZQ}, measured at different ν_{CPMG} (see Figs. 5.11 and 5.12) to determine the parameters of the exchange process. Following [47], we assumed the population of the excited state to be 2.4 % with a single timescale for all the residues. Experimental uncertainties in the extracted parameters, reported in Table 5.3, were estimated *via* a Monte Carlo approach.

The global fitting procedure yields a rate of the slow process $k_{\text{ex}}^{\text{slow}} = (400 \pm 70)\text{s}^{-1}$, in agreement with [47]. Therefore, we can conclude that our MQ CPMG experiments are reporting on the same folding-unfolding process of the $\alpha 3$ helix as the one reported by single-quantum ^{15}N CPMG experiments.

However, the exchange contributions, calculated using Eq. 3.4, along with the data in Table 5.3, the populations of the ground and excited states given in [47] and the above timescale, are rather small. Thus, we can conclude that the residues that show the highest values of μ_{MQ} must experience additional faster motions that have not been reported previously.

5.2.3 *Quantifying Chemical Exchange by HDR Relaxation Dispersion*

In order to characterize such motions, we measured a full HDR dispersion profile of L620 by using the pulse sequences of the Chap. 4. The data point at $\omega_1/(2\pi) = 0$ was obtained by subtracting to the cross-relaxation rate, measured by the K&K experiment [27], the exchange contribution of the slow process, calculated as described in the Sect. 5.2.2, and finally dividing the result by 2 to take into account the averaging of μ_{MQ} that can be explained by ALT. The dispersion profile is shown in Fig. 5.13. A small decrease of μ_{MQ} with increasing *rf* amplitude indicates the presence of exchange processes with rates comparable to the maximum *rf* amplitudes employed.

We extracted the parameters $k_{\text{ex}}^{\text{fast}} = (7,000 \pm 1,000)\text{s}^{-1}$ and $\overline{\Delta\omega}^2 = (18,000 \pm 5,000)\text{s}^{-2}$ by fitting the dispersion profile to Eq. 3.46.

Our results show that additional faster processes occur in the native state of KIX. It is worth noting that multiple-quantum relaxation is not sensitive to proton exchange with water molecules, which may introduce effects due to scalar relaxation in single-quantum experiments. Thus the faster processes that we identified and quantified are most likely due to conformational dynamics. Therefore, the folding of KIX from its intermediate state is not a simple two-state process and is best described in terms

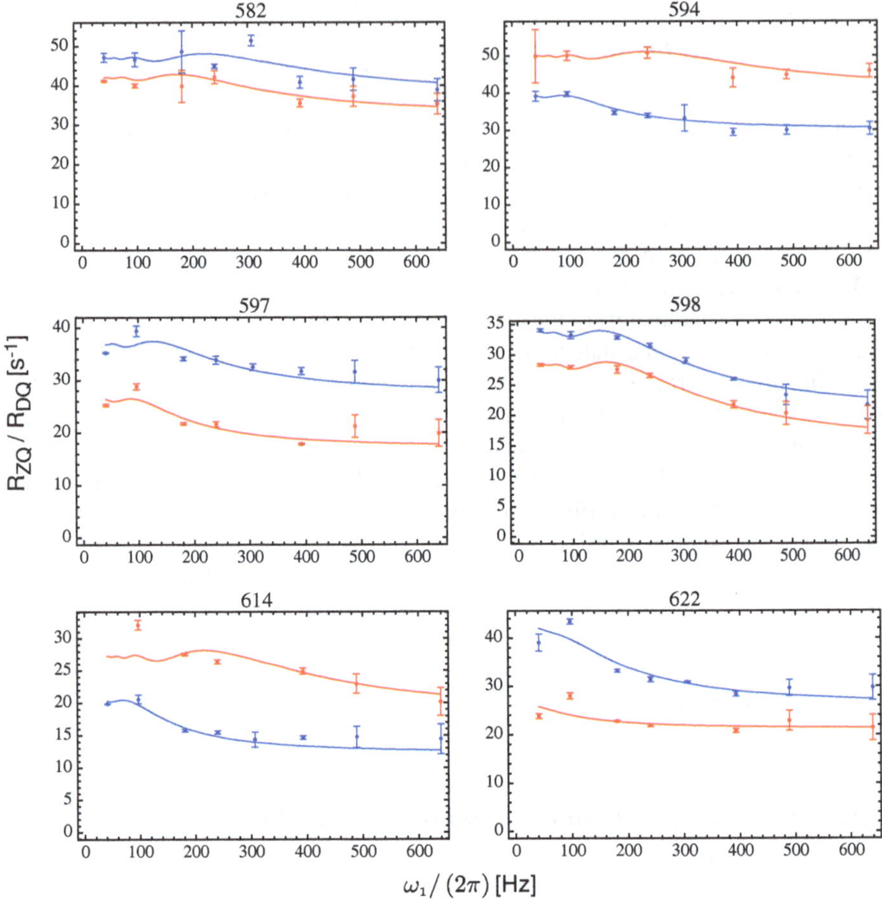

Fig. 5.11 Some MQ CPMG dispersion profiles in KIX. Zero- and double-quantum relaxation rates are marked by *red* and *blue dots*, respectively. The solid lines correspond to global fits to Eq. 3.11

of several consecutive steps that occur on different timescales and embrace different parts of the protein. Interestingly, these motions create conformational disorder in specific regions of the domain that are involved in the binding of the mixed lineage leukemia protein (MLL) [42] and KID [44]. Flexibility in these regions could possibly facilitate recognition and play a role in induced-fit mechanisms upon target binding [51].

5.3 Internal Dynamics in Engrailed 2

Homeoproteins constitute a large class of transcription factors present in many of eukaryotic species, from yeast to human. Homeoprotein Engrailed 2 plays a key

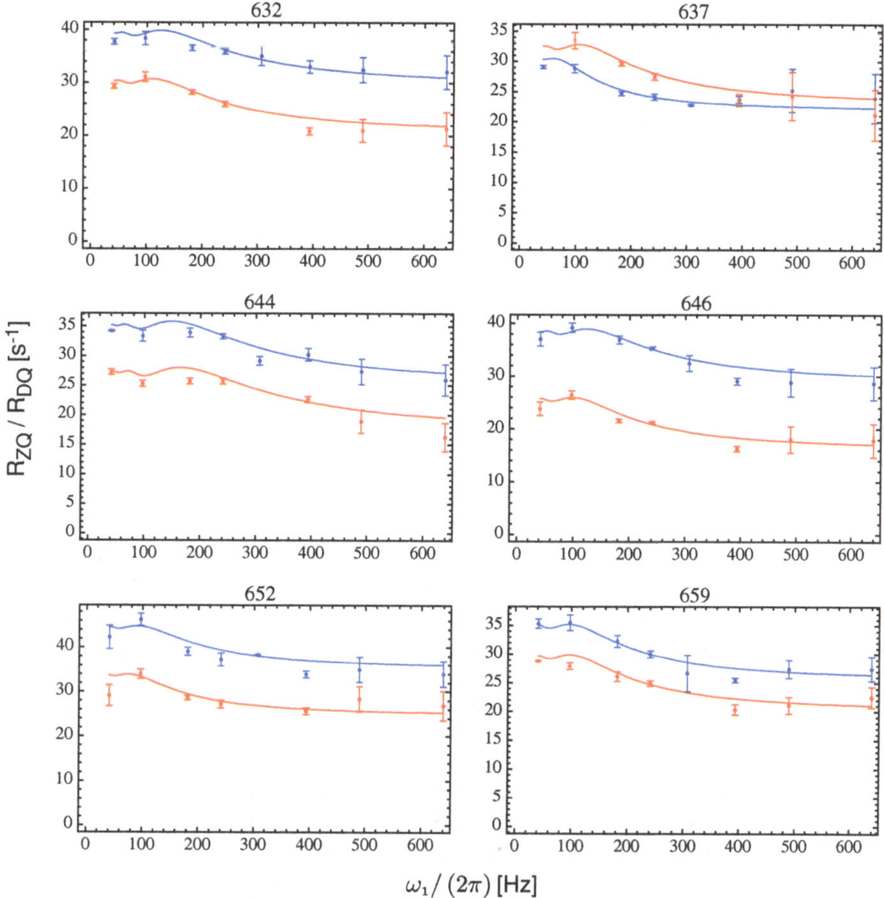

Fig. 5.12 Further examples of MQ CPMG dispersion profiles in KIX. Zero- and double-quantum relaxation rates are marked by *red* and *blue dots*, respectively. The solid lines correspond to global fits to Eq. 3.11

role in embryo development, regulating the differentiation and growth of the central nervous system [52], and in adult dopaminergic neurons, where the inhibition of the expression of Engrailed 2 causes apoptosis [53]. Chicken Engrailed 2 is a 289 residue protein. While the structure of the homeodomain (residues 200–259) is well known ([54], see Fig. 5.14), very little is known about the structure and dynamics of other parts of the protein.

We studied a construct of Engrailed 2 that comprises residues 146–259 (13.5 kDa), corresponding to the homeodomain with an N-terminal extension (Fig. 5.15). This intrinsically disordered extension contains several binding sites for other transcription factors (Pbx, FoxA2) and participates in the regulation of transcription [56, 57]. The different degrees of order in the two regions are evident in the HSQC spectrum in Fig. 5.16. Indeed, the NH resonances corresponding to homeodomain residues are

Table 5.3 Global fitting parameters of MQ CPMG dispersion profiles in KIX

Residue	$\Delta\omega_{DQ}$ [s^{-1}]	$\Delta\omega_{ZQ}$ [s^{-1}]	R^0_{DQ} [s^{-1}]	R^0_{ZQ} [s^{-1}]
582	4,100 ± 100	3,200 ± 100	37.4 ± 0.5	32.5 ± 0.5
585	6,700 ± 700	3,600 ± 700	8.8 ± 0.6	9.4 ± 0.6
586	9,700 ± 50	790 ± 10	2.8 ± 0.5	10.9 ± 0.1
588	1,300 ± 100	1,500 ± 200	12.2 ± 0.1	10.7 ± 0.1
590	480 ± 30	1,500 ± 200	27.9 ± 0.5	26.3 ± 0.9
593	4,600 ± 100	4,100 ± 300	13.8 ± 0.4	20.1 ± 0.4
594	1,800 ± 200	4,400 ± 600	30.2 ± 0.2	40.2 ± 0.4
597	2,400 ± 100	1,700 ± 300	27.5 ± 0.5	17.2 ± 0.2
598	2,500 ± 600	2,600 ± 700	24.6 ± 0.4	19.6 ± 0.4
599	3,300 ± 800	1,800 ± 300	27 ± 1	24.9 ± 0.2
601	1,500 ± 500	940 ± 200	28.2 ± 0.3	17.1 ± 0.1
612	2,500 ± 900	1,700 ± 400	7.8 ± 0.4	11.1 ± 0.3
616	1,600 ± 200	1,900 ± 200	17.3 ± 0.2	17.4 ± 0.1
620	2,500 ± 100	2,000 ± 800	10.2 ± 0.5	30.2 ± 0.3
624	2,000 ± 400	1,200 ± 500	44.4 ± 0.2	29.9 ± 0.2
626	2,000 ± 300	1,700 ± 300	27.0 ± 0.2	21.1 ± 0.2
627	2,000 ± 700	2,300 ± 700	23.6 ± 0.4	18.7 ± 0.4
629	2,700 ± 500	2,300 ± 600	28.1 ± 0.3	23.2 ± 0.3
630	1,700 ± 300	2,900 ± 600	23.8 ± 0.2	19 ± 1
632	2,400 ± 700	2,200 ± 500	29.9 ± 0.4	21.0 ± 0.3
633	2,700 ± 100	2,300 ± 800	26.7 ± 0.4	16.0 ± 0.4
634	2,200 ± 500	2,200 ± 600	24.3 ± 0.3	19.2 ± 0.4
636	2,300 ± 500	1,400 ± 500	27.1 ± 0.3	18.8 ± 0.3
638	1,500 ± 100	1,300 ± 800	25.5 ± 0.5	18.5 ± 0.3
640	4,300 ± 300	4,800 ± 100	26.9 ± 0.4	19.9 ± 0.5
641	2,700 ± 100	2,400 ± 400	25.3 ± 0.5	19.9 ± 0.3
645	4,300 ± 200	1,600 ± 300	21.8 ± 0.4	25.7 ± 0.6
646	2,400 ± 500	1,900 ± 300	29.0 ± 0.3	16.4 ± 0.2
651	2,300 ± 500	1,900 ± 300	28.1 ± 0.3	24.9 ± 0.2
652	1,800 ± 400	1,500 ± 200	35.5 ± 0.2	25.0 ± 0.1
655	3,400 ± 400	1,700 ± 200	21.3 ± 0.8	20.9 ± 0.2
671	1,200 ± 300	2,100 ± 400	10.8 ± 0.8	8.9 ± 0.3

well dispersed, while those belonging to the N-terminal extension all lie in a narrow range of about 1 ppm, which corfirms that the regions outside the homeodomain are intrinsically disordered.

In this section we use a combination of relaxation rates and magnetization transfer rates to provide a qualitative characterization of the internal dynamics in Engrailed 2. Also, ^{15}N $R_{1\rho}$ experiments are used to determine the timescale of such motions.

Expression and purification of the protein and sample preparation are described in [55, 58]. All experiments were carried out at 303 K at 18.79 T with a Bruker Avance 800 spectrometer equipped with a TXI cryoprobe with z-axis gradients.

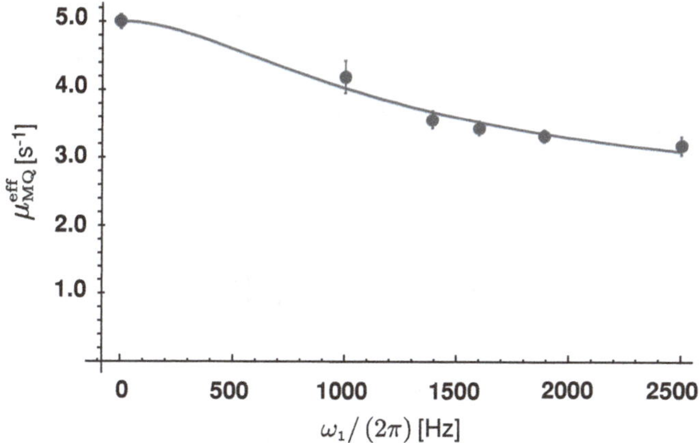

Fig. 5.13 HDR relaxation dispersion profiles for the backbone amide pair of L620 in KIX. The solid line correspond to a global fit to Eq. 3.46

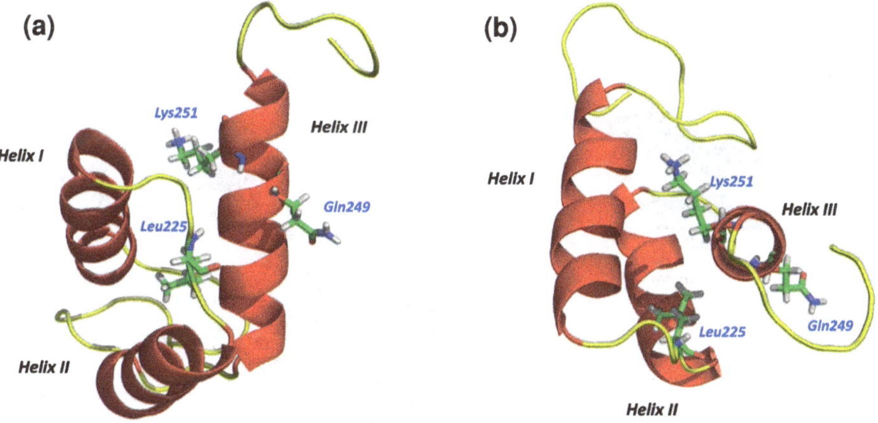

Fig. 5.14 Structure of chicken Engrailed 2 homeodomain solved by NMR spectroscopy [55]. Vertical (**a**) and horizontal (**b**) views of the protein with respect to the Helix III are represented. Residues whose mutations affect DNA binding are labelled

5.3.1 Identifying Chemical Exchange from the Relaxation of Single-Quantum Coherences

Chemical-exchange processes were identified by using a series of ^{15}N relaxation measurements, such as longitudinal ^{15}N relaxation rates (R_1), transverse ^{15}N relaxation rates under a single echo (R_2^{echo}) or under a Carr-Purcell-Meilboom-Gill train of echoes (R_2^{CPMG}), ^{15}N-^{1}H dipolar cross-relaxation rates (^{1}H-^{15}N NOE rate or σ_{NH}), and longitudinal (η_z) and transverse (η_{xy}) cross-relaxation rates resulting from the cross-correlation of the nitrogen-15 chemical shift anisotropy and the dipolar interaction between the nitrogen nucleus and its attached proton. The reader is referred to

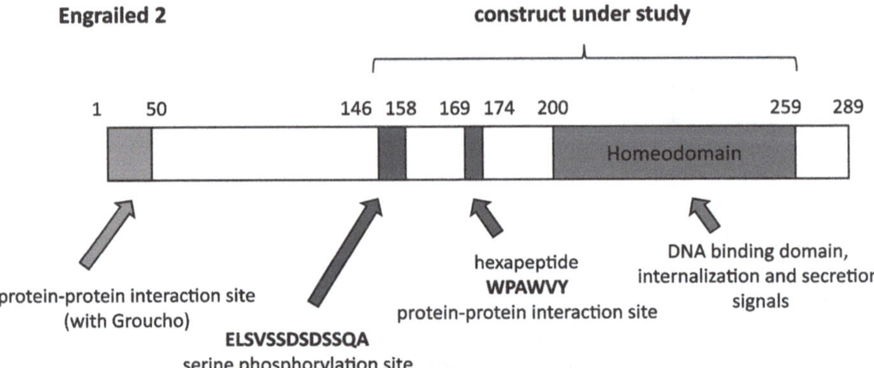

Fig. 5.15 Reproduced from [55]. Functionally important regions of Engrailed 2 homeoprotein. DNA binding domain (homeodomain), which contains also internalization and secretion signals is marked in *red*. The serine-rich fragment phosphorylated by CK2 kinase is marked in *blue*. Protein-protein interaction sites are marked in *orange* (Groucho protein) and in *green* (PBX and FoxA2 proteins)

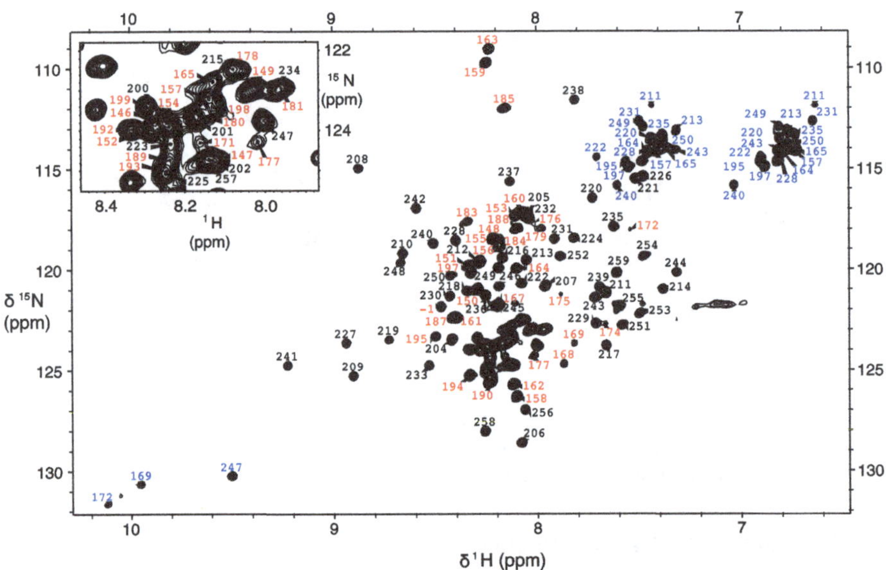

Fig. 5.16 Reproduced from [55, 58]. Assigned 2D ^{1}H-^{15}N HSQC spectrum of the ^{15}N-labelled 146–259 fragment of Engrailed 2 acquired at 500 MHz. A zoom of the central crowded region of the spectrum is shown in *upper-left corner*. Backbone resonances corresponding to residues in the homeodomain (200–259) and in the N-terminal extension (146–99) are marked in *black* and *red*, respectively. The side-chain resonances of Trp, Asn and Gln residues are indicated in *blue*

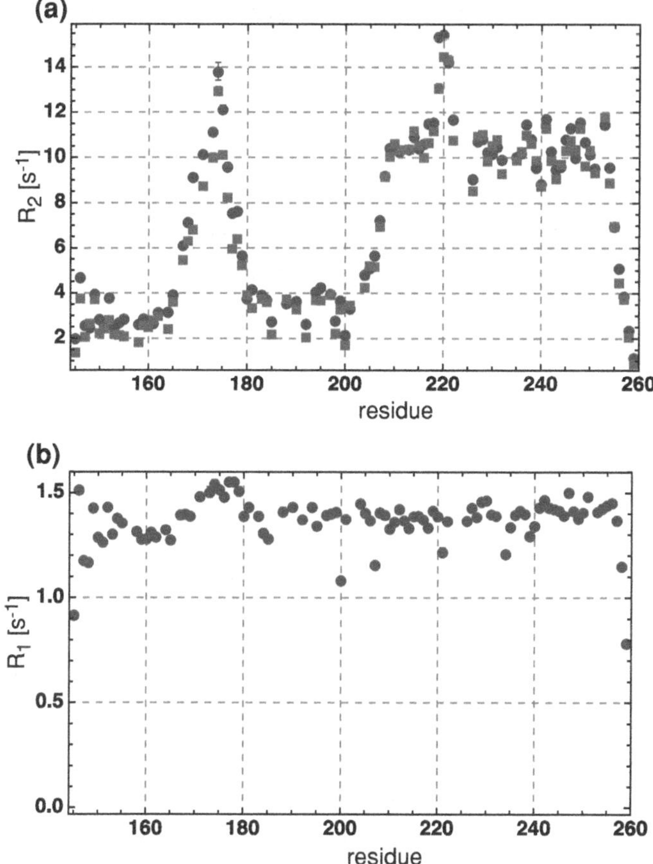

Fig. 5.17 Backbone amide transverse (**a**) and longitudinal (**b**) ^{15}N relaxation rates in Engrailed 2. In panel **a**, rates measured by refocusing the magnetization with a single echo are marked by *blue dots*, whereas rates measured by applying a CPMG train with a repetition rate of 1,000 Hz are marked by *red squares*

Sect. 4.1 for the pulse sequences and some references describing the details of these experimental techniques.

Transverse relaxation rates are shown in Fig. 5.17a. Many residues in the structured homeodomain, which is a DNA-binding globular domain, and residues 168–180 (hexapeptide region, which has been identified as a protein-protein interaction site) are characterized by remarkably high transverse relaxation rates. On the contrary, transverse relaxation in the rest of the unstructured region is much slower.

High relaxation rates can be due to the presence of structure [59], to chemical-exchange contributions, or both. R_2 rates measured under a CPMG train with a repetition rate of 1 kHz are only slightly lower than those measured under a single echo. Thus contributions of exchange processes in the ms time scale are small at best. Most exchange contributions to relaxation comes from sub-ms processes, as

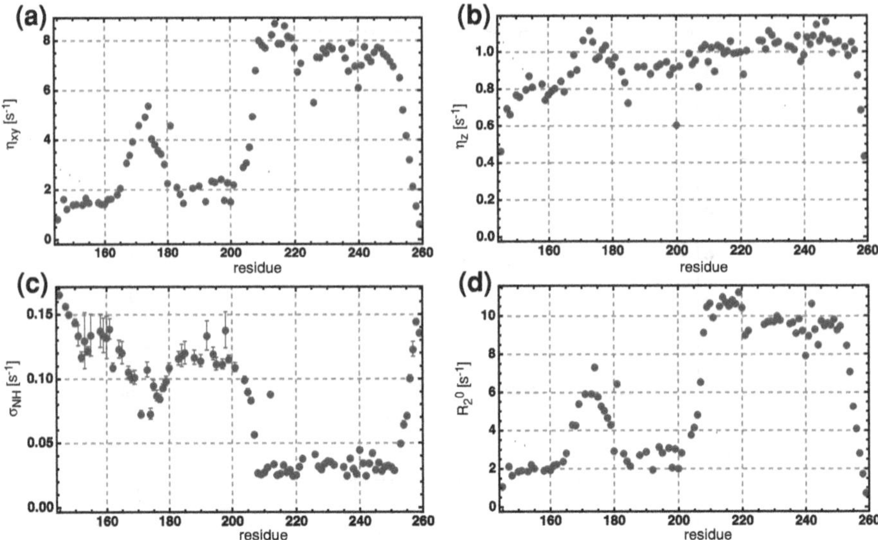

Fig. 5.18 N/NH CSA/DD transverse (**a**) and longitudinal (**b**) cross-relaxation rates, heteronuclear ^{15}N-^{1}H NOE values (**c**) and exchange-free ^{15}N transverse relaxation rates (**d**) in Engrailed 2

confirmed by the fact that no decrease in R_2 is observed in ^{15}N CPMG dispersion profiles (results not shown here).

Since exchange phenomena do not affect NOE rates to first order, they can provide more insights into the origin of such fast relaxation. As shown in Fig. 5.18c, most residues of the homeodomain are characterized by small σ_{NH} values, which confirms the rigidity of this domain, due to the structure that contains a bundle of three α-helices. On the other hand, much higher rates are found for the residues in the unstructured region. This is a clear signature of disorder [60–62].

Smaller NOE rates reveal restricted motions, which may be due to residual structure. This is the case for the hexapeptide, which has indeed a significant α-helical propensity [55].[1] Therefore, we can conclude that at least part of the fast transverse relaxation in the homeodomain and in the hexapeptide is due to residual structure.

Like σ_{NH}, η_{xy} is sensitive to local order and not to exchange processes, and therefore the values in Fig. 5.18a reflect the structure given by secondary elements. Higher rates

[1] The results of heteronuclear NOE measurements are commonly presented as the ratio of the signal intensities measured at the steady state under effective proton saturation (I^{ss}) and at equilibrium (I^{eq}). Such ratio is close to 1 for ordered regions and lower for disordered ones, and depends on both σ_{NH} and R_1 [63]:

$$\frac{I^{ss}}{I^{eq}} = 1 + \frac{\gamma_H \sigma_{NH}}{\gamma_N R_1}.$$

In the case of the hexapeptide region σ_{NH} is smaller than the values measured in the rest of the N-terminal extension, while R_1 is larger. Therefore, exploiting the combined effects on σ_{NH} and R_1, I^{ss}/I^{eq} is more sensitive to the presence of order than the two probes individually.

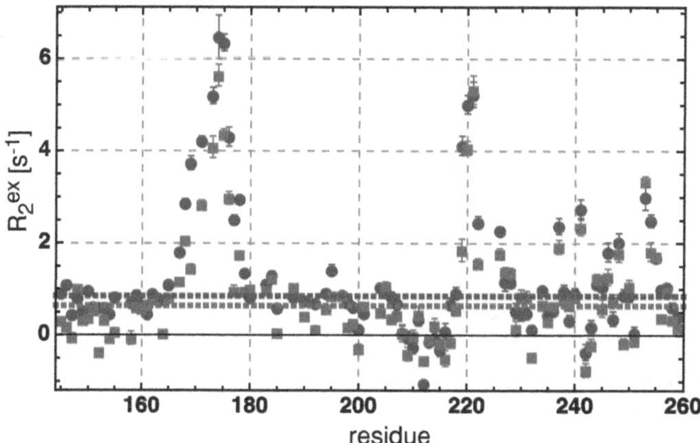

Fig. 5.19 Contributions of chemical exchange to the backbone amide transverse ^{15}N relaxation rates in Engrailed 2. Rates measured by refocusing the magnetization with a single echo are marked by *blue dots*, whereas rates measured by applying a CPMG train with a repetition rate of 1,000 Hz are marked by *red squares*. The threshold values, marked by *blue* and *red dotted lines* for single-echo and CPMG experiments, respectively, were obtained by computing the average of the exchange contributions excluding the 25 highest and lowest values

are obtained in regions that have been identified as ordered because of the NOE rates, while lower values are obtained in more flexible areas.

A combination of η_{xy}, η_z, σ_{NH} and R_1 rates yields an estimate of the exchange-free relaxation rates (Fig. 5.18d) that are only due to CSA and DD relaxation, according to the expression of Kroenke et al. [64]:

$$R_2^0 = (R_1 - 1.249\sigma_{NH})\,\frac{\eta_{xy}}{\eta_z} + 1.079\sigma_{NH}. \tag{5.1}$$

The values of the chemical exchange contributions to transverse relaxation in Fig. 5.19 are simply given by

$$R_2^{ex} = R_2 - R_2^0. \tag{5.2}$$

Significant exchange contributions are identified for the residues whose values lie above the threshold in Fig. 5.19. Thus, fluctuations at sub-ms timescales are likely to occur both in the hexapeptide and in a few restricted locations in the homeodomain, such as residues 219–221.

5.3.2 *Quantifying Chemical Exchange by $R_{1\rho}$ Relaxation Dispersion*

Such motions have been characterized using ^{15}N $R_{1\rho}$ relaxation dispersion NMR. The pulse sequences are described in Sect. 4.1. Experimental data for nine residues identi-

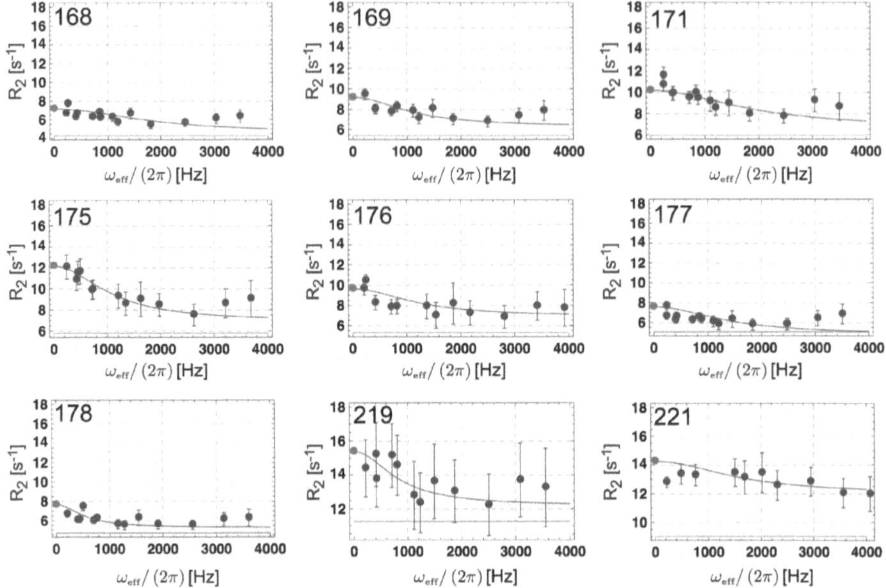

Fig. 5.20 Representative $R_{1\rho}$ relaxation dispersion profiles in Engrailed 2. Experiments were carried on the backbone amide ^{15}N of the residues indicated by the labels in the panels. The corresponding exchange-free ^{15}N transverse relaxation rates are marked by *red solid lines*. *Light blue dots* indicate the values of R_2 measured with a single echo

fied as outliers in Fig. 5.19 are reported in Fig. 5.20. The fitting of the experimental data to Eq. 3.14 using the genetic algorithm of Chap. 4 and Appendix C yields the parameters reported in Table 5.4.

With the exception of 178, all residues in the hexapeptide region feature kinetic rates of the same order of magnitude. Nevertheless, preliminary attempts to perform a global fit to extract a single timescale for residues 168–177 were not successful. The residual exchange contribution $R_2^{0,\mathrm{fit}} - R_2^{0,\mathrm{exp}}$ is not significant (i.e., smaller than the error bar on $R_2^{0,\mathrm{fit}}$) for all residues in the hexapeptide region except 175–176.

This observation suggest that μs motions faster than the maximum $\omega_{\mathrm{eff}}/(2\pi)$ used in our experiments ($\approx$ 4 kHz) are not likely to occur. Thus, the following interpretations of the relaxation dispersion data could be proposed:

1. Complex motions occur on different timescales, even in consecutive residues. These must therefore be completely uncorrelated;
2. Motions involving more than two sites occur. This would require a change of the model used to fit the data;
3. The error bars on the timescale extracted from the fit of the dispersion profiles are underestimated, so that the different kinetic rates are in fact compatible with a single process.

More experimental data, possibly acquired at different B_0 field strengths, are required to determine the correct interpretation unambiguously. One attractive option is to

Table 5.4 Fitting parameters extracted from ^{15}N $R_{1\rho}$ dispersion profiles in Engrailed 2

Residue	k_{ex} [s^{-1}]	$p_A p_B \Delta\omega^2$ [s^{-2}]	$R_2^{0,fit}$ [s^{-1}]	$R_2^{0,fit} - R_2^{0,exp}$ [s^{-1}]
168	11,400 ± 400	30,000 ± 3,000	4.6 ± 0.7	0.3
169	7,000 ± 200	21,000 ± 1,000	6.3 ± 0.9	0.8
171	11,700 ± 200	41,000 ± 2,000	7 ± 1	0.7
175	6,800 ± 200	36,000 ± 1,000	7 ± 1	1.1
176	7,600 ± 300	21,000 ± 3,000	7 ± 1	1.6
177	8,600 ± 800	24,000 ± 5,000	4.8 ± 0.7	0.3
178	3,400 ± 200	8,000 ± 400	5.3 ± 0.8	0.6
219	5,200 ± 700	17,000 ± 3,000	12 ± 2	0.9
221	10,000 ± 1,000	23,000 ± 6,000	12 ± 2	3.0

decrease the temperature. This would slow down the exchange processes, possibly as slow as the ms timescale. This could be easily investigated by ^{15}N CPMG relaxation dispersion. Indeed CPMG experiments require much less experimental time than $R_{1\rho}$ experiments, allowing one to record full dispersion profiles with more data points than $R_{1\rho}$ and at different static magnetic fields even with a single protein sample. These advantages are of primary importance for systems with limited stability such as Engrailed.

Nevertheless, it is evident that the hexapeptide motif is characterized by complex dynamics, which is likely to be linked with its physiological function. Moreover, the time scale of motions in the homeodomain is reasonably close to that of the hexapeptide. It is therefore tempting to suppose that the dispersion of residue 219–221 is due to transient contacts with the unstructured regions, and in particular with the hexapeptide. Long-range interactions between the two regions has indeed been identified [55]. This hypothesis could be verified by studying the dynamics of different constructs, for instance containing only the homeodomain or only the N-terminal extension.

References

1. Schneider DM, Dellwo MJ, Wand AJ (1992) Biochemistry 31:3645
2. Tjandra N, Feller SE, Pastor RW (1995) J Am Chem Soc 117:12562
3. Showalter SA, Brüschweiler R (2007) J Chem Theory Comput 3:961
4. Maragakis P, Lindorff-Larsen K (2008) J Phys Chem B 112:6155
5. Massi F, Grey MJ, Palmer AG (2005) Protein Sci 14:735
6. Majumdar A, Ghose R (2004) J Biomol NMR 28:213
7. Pelupessy P, Ferrage F, Bodenhausen G (2007) J Chem Phys 126:134508
8. Hansen DF, Feng H, Zhou Z, Bai Y (2009) J Am Chem Soc 131:16257
9. Mills JL, Szyperski T (2002) J Biomol NMR 23:63
10. Dittmer J, Bodenhausen G (2004) J Am Chem Soc 126:1314
11. Wist J, Frueh D, Tolman JR, Bodenhausen G (2004) J Biomol NMR 28:263
12. Ban D, Funk M, Gulich R, Egger D, Sabo TM, Walter KFA, Fenwick RB, Giller K, Pichierri F, de Groot BL, Lange OF, Grubmüller H, Salvatella X, Wolf M, Loidl A, Kree R, Becker S, Lakomek N-A, Lee D, Lunkenheimer P, Griesinger C (2011) Angew Chem Int Ed 50:11437

13. Peti W, Meiler J, Brüschweiler R, Griesinger C (2002) J Am Chem Soc 124:5822
14. Lakomek N-A, Walter KFA, Farès C, Lange OF, Groot BL, Grubmüller H, Brüschweiler R, Munk A, Becker S, Meiler J, Griesinger C (2008) J Biomol NMR 41:139
15. Salmon L, Bouvignies G, Markwick P, Blackledge M (2011) Biochemistry 50:2735
16. Lange OF, Lakomek NA, Farès C, Schröder GF, Walter KFA, Becker S, Meiler J, Grubmuller H, Griesinger C, de Groot BL (2008) Science 320:1471
17. Fenwick RB, Esteban-Martín S, Richter B, Lee D, Walter KFA, Milovanovic D, Becker S, Lakomek N-A, Griesinger C, Salvatella X (2011) J Am Chem Soc 133:10336
18. Penengo L, Mapelli M, Murachelli AG, Confalonieri S (2006) Cell 124:1183
19. Lee S, Tsai YC, Mattera R, Smith WJ, Kostelansky MS, Weissman AM, Bonifacino JS, Hurley JH (2006) Nature Struct Mol Biol 13:264
20. Eddins MJ, Carlile CM, Gomez KM, Pickart CM, Wolberger C (2006) Nature Struct Mol Biol 13:915
21. Dikic I, Wakatsuki S, Walters KJ (2009) Nature Rev Mol Cell Biol 10:659
22. Cornilescu G, Marquardt JL, Ottiger M, Bax A (1998) J Am Chem Soc 120:6836
23. Schrödinger LLC (2010) The PyMOL molecular graphics system, version 1.3r1
24. Wang C, Palmer AG (2003) Magn Reson Chem 41:866
25. Pelupessy P, Espargallas GM, Bodenhausen G (2003) J Magn Reson 161:258
26. Salvi N, Ulzega S, Ferrage F, Bodenhausen G (2012) J Am Chem Soc 134:2481
27. Kloiber K, Konrat R (2000) J Biomol NMR 18:33
28. Cole R, Loria JP (2002) Biochemistry 41:6072
29. Doucet N, Khirich G, Kovrigin EL, Loria JP (2011) Biochemistry 50:1723
30. Sidhu A, Surolia A, Robertson AD, Sundd M (1037) J Mol Biol 2011:411
31. Huang KY, Amodeo GA, Tong L, McDermott A (2011) Protein Sci 20:630
32. Igumenova TI, Wand AJ, McDermott AE (2004) J Am Chem Soc 126:5323
33. Zandarashvili L, Li D-W, Wang T, Brüschweiler R, Iwahara J (2011) J Am Chem Soc 133:9192
34. Esadze A, Li D-W, Wang T, Brüschweiler R, Iwahara J (2011) J Am Chem Soc 133:909
35. Vögeli B, Segawa TF, Leitz D, Sobol A (2009) J Am Chem Soc 131:17215
36. Kitahara R, Yokoyama S, Akasaka K (2005) J Mol Biol 347:277
37. Meiler J, Peti W, Griesinger C (2003) J Am Chem Soc 125:8072
38. Gunasekaran K, Ma B, Nussinov R (2004) Proteins Struct Funct Bioinf 57:433
39. Long D, Brüschweiler R (2011) J Am Chem Soc 133:18999
40. Garner TP, Strachan J, Shedden EC, Long JE, Cavey JR, Shaw B, Layfield R, Searle MS (2011) Biochemistry 50:9076
41. Radhakrishnan I, Perez-Alvarado GC, Parker D, Dyson HJ, Montminy MR, Wright PE (1997) Cell 91:741
42. Goto NK, Zor T, M-Yamout M, Dyson HJ, Wright PE (2002) J Biol Chem 277:43168
43. Eliezer D, Palmer AG (2007) Nature 447:920
44. Sugase K, Dyson HJ, Wright PE (1021) Nature 2007:447
45. Horng J-C, Tracz SM, Lumb KJ, Raleigh DP (2005) Biochemistry 44:627
46. Tollinger M, Kloiber K, Agoston B, Dorigoni C, Lichtenecker R, Schmid W, Konrat R (2006) Biochemistry 45:8885
47. Schanda P, Brutscher B, Konrat R, Tollinger M (2008) J Mol Biol 380:726
48. Radhakrishnan I, Pérez-Alvarado GC, Parker D, Dyson HJ, Montminy MR, Wright PE (1997) Cell 91:741
49. Orekhov VY, Korzhnev DM, Kay LE (1886) J Am Chem Soc 2004:126
50. Korzhnev DM, Kloiber K, Kay LE (2004) J Am Chem Soc 126:7320
51. Mittag T, Kay LE, Forman-Kay JD (2010) J Mol Recognit 23:105
52. Joyner AL (1996) Trends Genet 12:15
53. McGinnis W, Levine MS, Hafen E, Kuroiwa A (1984) Nature 308:428
54. Fraenkel E, Rould MA, Chambers KA, Pabo CO (1998) J Mol Biol 284:351
55. Augustyniak R (2011) NMR studies of the partially disordered protein Engrailed 2 and new NMR methods for diffusion measurements and protein sidechain assignments. Ph.D. thesis, Université Pierre et Marie Curie

56. Foucher I, Montesinos ML, Volovitch M, Prochiantz A, Trembleau A (1867) Development 2003:130
57. Piper DE, Batchelor AH, Chang CP, Cleary ML, Wolberger C (1999) Cell 96:587
58. Augustyniak R, Balayssac S, Ferrage F, Bodenhausen G, Lequin O (2011) Biomol NMR Assign 5:229
59. Klein-Seetharaman J, Oikawa M, Grimshaw SB (2002) Science 295:1657
60. Buevich AV, Shinde UP, Inouye M, Baum J (2001) J Biomol NMR 20:233
61. Mackay JP, Muiznieks LD, Toonkool P, Weiss AS (2005) J Struct Biol 150:154
62. Prasch S, Schwarz S, Eisenmann S, Wöhrl BM, Schweimer K, Rösch P (2006) Biochemistry 45:4542
63. Ferrage F (2012) Methods Mol Biol 831:141
64. Kroenke CD, Loria JP, Lee LK, Rance M (1998) J Am Chem Soc 120:7905

Chapter 6
Conclusions

Protein function often depends on motions and conformational rearrangements occurring on a μs-ms time scale. Such processes induce chemical exchange effects and are effectively characterized by relaxation dispersion, i.e., the variation of relaxation rates as a function of the amplitude of an *rf* field or as a function of the repetition rate of π pulses. The goal of the present work was to develop analytical tools to understand and control the chemical-exchange induced multiple-quantum cross-relaxation under our heteronuclear double resonance (HDR) using a wideband alternating-phase low-power technique for zero-residual-splitting (WALTZ)-32 sequence and to apply this method, in combination with conventional ones, such as single- and multiple-quantum relaxation rates measured by using Carr Purcell Meiboom Gill (CPMG) and $R_{1\rho}$ experiments, to the characterization of internal motions in proteins.

A fully analytical study of multiple-quantum cross-relaxation dispersion during an HDR WALTZ-32 pulse sequence in the framework of Redfield's theory is presented. We derived a compact analytical expression, valid for fast exchange between two conformers, that describes the dependence of the effective cross-relaxation rate μ_{MQ} on the applied *rf* amplitude ω_1, on the rate of the exchange process k_{ex}, and on a parameter $\overline{\Delta\omega}^2$ defined as the product of the populations of the two sites and of the chemical shift differences between the sites for the two spins. Formally, the expression is analogous to the well-known $R_{1\rho}$ relaxation dispersion for single-quantum coherences, modified by a correction factor that takes into account the phase modulation inherent to HDR WALTZ-32 pulse sequences. Numerical simulations and experiments were used to confirm the validity of our expression.

Our expression allowed us to use the HDR method as a new tool to probe local dynamics that occur on microsecond timescales and to characterize correlated exchange processes in molecules of biological interest through the investigation of multiple-quantum cross relaxation. Indeed, it is possible to extract quantitative information about kinetics (i.e., exchange rates k_{ex}) from the observed relaxation dispersion. In combination with other techniques, e.g. $R_{1\rho}$ relaxation dispersion, our analytical formula also allows one to extract information about the thermodynamics (i.e., populations of the two sites) and structural parameters (chemical shift

N. Salvi, *Dynamic Studies Through Control of Relaxation in NMR Spectroscopy*, Springer Theses, DOI: 10.1007/978-3-319-06170-2_6,

differences of the two spins), thus permitting an accurate and detailed quantification of local dynamic processes.

We used multiple-quantum relaxation techniques to identify the extent of chemical exchange in human ubiquitin. Several regions of ubiquitin show the signature of chemical exchange, including the hydrophobic patch and the $\beta 4$-$\alpha 2$ loop, which are both involved in many interactions. The heteronuclear double resonance method was used to determine the time scales of motions that give rise to chemical exchange. Dispersion profiles were obtained for the backbone NH^N pairs of several residues in the hydrophobic patch and the $\beta 4$-$\alpha 2$ loop, as well as the C-terminus of helix $\alpha 1$. We show that a single time scale (ca. 50 μs) can be used to fit the dispersion profiles for most residues. Potential mechanisms for the propagation of motions and the possible extent of correlation of these motions were discussed.

Multiple-quantum relaxation techniques were also employed to explore the internal dynamics in the KID-binding (KIX) domain of the CREB-binding protein (CBP). Evident signs of chemical exchange in KIX were found in the cross-relaxation rates of multiple-quantum coherences, especially in the N-terminus and in the loop between the helices $\alpha 1$ and $\alpha 2$. Multiple-quantum CPMG were used to characterize a slow folding-unfolding process ($k_{ex} \approx 400\,s^{-1}$) that was already characterized by Konrat and coworkers. Our results pointed out that the slow process is only marginally responsible for the chemical exchange contributions to cross-relaxation rates of multiple-quantum coherences. Thus faster processes must be present. We used the HDR method to quantify the time scale of the fast process for L620, $k_{ex} = (7 \pm 1) \times 10^3\,s^{-1}$. Additional work is required to probe the time scale of fast processes for other residues.

We also studied a construct of Engrailed 2 that comprises the folded homeodomain and an intrinsically disordered N-terminal extension, containing several binding sites for other transcription factors. Among these binding sites an important role is played by the hexapeptide 169–174, which is a protein-protein interaction site. A combination of ^{15}N relaxation rates and magnetization transfer rates were used to obtain a qualitative characterization of internal dynamics in Engrailed 2. While contributions of exchange processes on the ms time scale were small, fluctuations at sub-ms timescales were found to occur both in the hexapeptide and in a few restricted locations in the homeodomain, such as residues 219–221. Such motions were characterized using ^{15}N $R_{1\rho}$ relaxation dispersion NMR. For most residues in the hexapeptide region, kinetic rates of the same order of magnitude $\approx 7000\,s^{-1}$ were extracted from the dispersion profiles. However preliminary attempts to perform a global fit to extract a single time scale for residues in this region were not successful. This failure could reflect the fact that motions of close or even consecutive residues do actually occur on different time scales, or it could be the signature of the presence of motions that involve more than two sites, or it could be due to the limited extent of our data sets. More experimental data, possibly acquired at different static magnetic field strengths, are required to improve the interpretation of our data. One attractive option is to decrease the temperature at which the experiments are carried out. This would slow down the exchange processes, possibly up to a ms timescale that could be easily investigated by ^{15}N CPMG relaxation dispersion. Nevertheless, it is evident that the hexapeptide motif is characterized by complex dynamics, which may be linked to its physiological function. Moreover, the time scale of motions in the homeodomain is reasonably close to that of the hexapeptide. It is therefore tempting to suppose that the dispersion of residues 219–221 is due to transient contacts with the unstructured regions, and in particular with the hexapeptide. Long-range interactions between the two has indeed been identified and this hypothesis could be verified by studying the dynamics of different constructs, for instance containing only the homeodomain or only the N-terminal extension.

Appendix A
Explicit Forms of the Functions Approximated in Section 3.4

Henceforth we shall use $\alpha = \omega_1 \tau_{\text{ex}}$. Algebraic manipulations carried out with Mathematica 8 show that

$$
\begin{aligned}
\mathbb{F}_1(\alpha) = & -24 + 3e^{\frac{30\pi}{\alpha}} - 3e^{\frac{36\pi}{\alpha}} + 3e^{\frac{42\pi}{\alpha}} - 3e^{\frac{48\pi}{\alpha}} + 3e^{\frac{78\pi}{\alpha}} - 3e^{\frac{84\pi}{\alpha}} + 3e^{\frac{90\pi}{\alpha}} \\
& + 192e^{\frac{96\pi}{\alpha}} + \left(-3e^{\frac{3\pi}{2\alpha}} - 3e^{\frac{51\pi}{2\alpha}} + 2e^{\frac{63\pi}{2\alpha}} - 3e^{\frac{75\pi}{2\alpha}} - 2e^{\frac{87\pi}{2\alpha}} - 3e^{\frac{99\pi}{2\alpha}} + 4e^{\frac{103\pi}{2\alpha}} \right. \\
& - 2e^{\frac{105\pi}{2\alpha}} - e^{\frac{109\pi}{2\alpha}} + 4e^{\frac{115\pi}{2\alpha}} - 2e^{\frac{117\pi}{2\alpha}} + 4e^{\frac{127\pi}{2\alpha}} - 2e^{\frac{129\pi}{2\alpha}} - e^{\frac{133\pi}{2\alpha}} + 4e^{\frac{139\pi}{2\alpha}} \\
& - 2e^{\frac{141\pi}{2\alpha}} - 3e^{\frac{147\pi}{2\alpha}} - 4e^{\frac{151\pi}{2\alpha}} + 2e^{\frac{153\pi}{2\alpha}} + e^{\frac{157\pi}{2\alpha}} + 2e^{\frac{159\pi}{2\alpha}} - 4e^{\frac{163\pi}{2\alpha}} + 2e^{\frac{165\pi}{2\alpha}} \\
& \left. - 3e^{\frac{171\pi}{2\alpha}} - 4e^{\frac{175\pi}{2\alpha}} + 2e^{\frac{177\pi}{2\alpha}} + e^{\frac{181\pi}{2\alpha}} - 2e^{\frac{183\pi}{2\alpha}} - 4e^{\frac{187\pi}{2\alpha}} + 2e^{\frac{189\pi}{2\alpha}} \right) \alpha.
\end{aligned}
$$

The function $\mathbb{F}_2(\omega_1, \tau_{\text{ex}})$ can be written as $\mathbb{F}_2(\alpha) = A + B\alpha + C\alpha^2$, with

$$
\begin{aligned}
A = & \, 6 - 2e^{\frac{\pi}{\alpha}} + 6e^{\frac{3\pi}{\alpha}} - 5e^{\frac{4\pi}{\alpha}} - 2e^{\frac{9\pi}{2\alpha}} + e^{\frac{5\pi}{\alpha}} - e^{\frac{11\pi}{2\alpha}} + 10e^{\frac{6\pi}{\alpha}} - e^{\frac{13\pi}{2\alpha}} + 5e^{\frac{7\pi}{\alpha}} \\
& + 2e^{\frac{15\pi}{2\alpha}} + 3e^{\frac{8\pi}{\alpha}} + 2e^{\frac{17\pi}{2\alpha}} + 6e^{\frac{10\pi}{\alpha}} - 3e^{\frac{21\pi}{2\alpha}} - e^{\frac{11\pi}{\alpha}} - e^{\frac{23\pi}{2\alpha}} + 15e^{\frac{12\pi}{\alpha}} + e^{\frac{25\pi}{2\alpha}} \\
& + 5e^{\frac{13\pi}{\alpha}} + e^{\frac{27\pi}{2\alpha}} + 2e^{\frac{14\pi}{\alpha}} + 6e^{\frac{15\pi}{\alpha}} - e^{\frac{31\pi}{2\alpha}} - 5e^{\frac{16\pi}{\alpha}} - e^{\frac{33\pi}{2\alpha}} + e^{\frac{17\pi}{\alpha}} - 2e^{\frac{35\pi}{2\alpha}} \\
& - 16e^{\frac{18\pi}{\alpha}} + 2e^{\frac{37\pi}{2\alpha}} - 3e^{\frac{19\pi}{\alpha}} + 5e^{\frac{39\pi}{2\alpha}} + 7e^{\frac{20\pi}{\alpha}} + 4e^{\frac{41\pi}{2\alpha}} + 11e^{\frac{21\pi}{\alpha}} - 10e^{\frac{43\pi}{2\alpha}} \\
& + 10e^{\frac{22\pi}{\alpha}} - 7e^{\frac{45\pi}{2\alpha}} - 2e^{\frac{23\pi}{\alpha}} - 2e^{\frac{47\pi}{2\alpha}} + 17e^{\frac{24\pi}{\alpha}} + 2e^{\frac{49\pi}{2\alpha}} + 8e^{\frac{25\pi}{\alpha}} + 2e^{\frac{51\pi}{2\alpha}} \\
& + 4e^{\frac{26\pi}{\alpha}} + 2e^{\frac{27\pi}{\alpha}} + 2e^{\frac{55\pi}{2\alpha}} - 3e^{\frac{28\pi}{\alpha}} + e^{\frac{29\pi}{\alpha}} - 3e^{\frac{59\pi}{2\alpha}} + 14e^{\frac{30\pi}{\alpha}} - 3e^{\frac{61\pi}{2\alpha}} + 5e^{\frac{31\pi}{\alpha}} \\
& + 8e^{\frac{63\pi}{2\alpha}} + 11e^{\frac{32\pi}{\alpha}} + 6e^{\frac{65\pi}{2\alpha}} + 24e^{\frac{33\pi}{\alpha}} - 4e^{\frac{67\pi}{2\alpha}} + 10e^{\frac{34\pi}{\alpha}} - 11e^{\frac{69\pi}{2\alpha}} - 3e^{\frac{35\pi}{\alpha}} \\
& - 3e^{\frac{71\pi}{2\alpha}} + 7e^{\frac{36\pi}{\alpha}} + 3e^{\frac{73\pi}{2\alpha}} + 15e^{\frac{37\pi}{\alpha}} + 3e^{\frac{75\pi}{2\alpha}} + 6e^{\frac{38\pi}{\alpha}} - 3e^{\frac{79\pi}{2\alpha}} - 3e^{\frac{40\pi}{\alpha}} \\
& - 3e^{\frac{81\pi}{2\alpha}} + e^{\frac{41\pi}{\alpha}} - 4e^{\frac{83\pi}{2\alpha}} - 6e^{\frac{42\pi}{\alpha}} + 4e^{\frac{85\pi}{2\alpha}} - 11e^{\frac{43\pi}{\alpha}} + 11e^{\frac{87\pi}{2\alpha}} + 15e^{\frac{44\pi}{\alpha}} \\
& + 8e^{\frac{89\pi}{2\alpha}} + 40e^{\frac{45\pi}{\alpha}} - 22e^{\frac{91\pi}{2\alpha}} + 18e^{\frac{46\pi}{\alpha}} - 15e^{\frac{93\pi}{2\alpha}} - 4e^{\frac{47\pi}{\alpha}} - 4e^{\frac{95\pi}{2\alpha}} + 31e^{\frac{48\pi}{\alpha}} \\
& + 4e^{\frac{97\pi}{2\alpha}} + 18e^{\frac{49\pi}{\alpha}} + 4e^{\frac{99\pi}{2\alpha}} + 8e^{\frac{50\pi}{\alpha}} + 6e^{\frac{51\pi}{\alpha}} + 4e^{\frac{103\pi}{2\alpha}} - 3e^{\frac{52\pi}{\alpha}} + 2e^{\frac{105\pi}{2\alpha}}
\end{aligned}
$$

N. Salvi, *Dynamic Studies Through Control of Relaxation in NMR Spectroscopy*, Springer Theses, DOI: 10.1007/978-3-319-06170-2,

$$
\begin{aligned}
&+5e^{\frac{53\pi}{\alpha}}-5e^{\frac{107\pi}{2\alpha}}+4e^{\frac{54\pi}{\alpha}}-5e^{\frac{109\pi}{2\alpha}}-3e^{\frac{55\pi}{\alpha}}+14e^{\frac{111\pi}{2\alpha}}+21e^{\frac{56\pi}{\alpha}}+10e^{\frac{113\pi}{2\alpha}}\\
&+50e^{\frac{57\pi}{\alpha}}-8e^{\frac{115\pi}{2\alpha}}+20e^{\frac{58\pi}{\alpha}}-19e^{\frac{117\pi}{2\alpha}}-9e^{\frac{59\pi}{\alpha}}-5e^{\frac{119\pi}{2\alpha}}+31e^{\frac{60\pi}{\alpha}}+5e^{\frac{121\pi}{2\alpha}}\\
&+25e^{\frac{61\pi}{\alpha}}+9e^{\frac{123\pi}{2\alpha}}+14e^{\frac{62\pi}{\alpha}}+6e^{\frac{63\pi}{\alpha}}-5e^{\frac{127\pi}{2\alpha}}-3e^{\frac{64\pi}{\alpha}}-7e^{\frac{129\pi}{2\alpha}}+11e^{\frac{65\pi}{\alpha}}\\
&-6e^{\frac{131\pi}{2\alpha}}-6e^{\frac{66\pi}{\alpha}}+6e^{\frac{133\pi}{2\alpha}}-13e^{\frac{67\pi}{\alpha}}+15e^{\frac{135\pi}{2\alpha}}+27e^{\frac{68\pi}{\alpha}}+12e^{\frac{137\pi}{2\alpha}}+63e^{\frac{69\pi}{\alpha}}\\
&-34e^{\frac{139\pi}{2\alpha}}+28e^{\frac{70\pi}{\alpha}}-23e^{\frac{141\pi}{2\alpha}}-14e^{\frac{71\pi}{\alpha}}-6e^{\frac{143\pi}{2\alpha}}+45e^{\frac{72\pi}{\alpha}}+6e^{\frac{145\pi}{2\alpha}}+28e^{\frac{73\pi}{\alpha}}\\
&+14e^{\frac{147\pi}{2\alpha}}+20e^{\frac{74\pi}{\alpha}}+26e^{\frac{75\pi}{\alpha}}+6e^{\frac{151\pi}{2\alpha}}-5e^{\frac{76\pi}{\alpha}}+8e^{\frac{153\pi}{2\alpha}}-15e^{\frac{77\pi}{\alpha}}-7e^{\frac{155\pi}{2\alpha}}\\
&-7e^{\frac{157\pi}{2\alpha}}-11e^{\frac{79\pi}{\alpha}}+16e^{\frac{159\pi}{2\alpha}}+25e^{\frac{80\pi}{\alpha}}+14e^{\frac{161\pi}{2\alpha}}+54e^{\frac{81\pi}{\alpha}}-12e^{\frac{163\pi}{2\alpha}}\\
&+28e^{\frac{82\pi}{\alpha}}-27e^{\frac{165\pi}{2\alpha}}-3e^{\frac{83\pi}{\alpha}}-7e^{\frac{167\pi}{2\alpha}}+27e^{\frac{84\pi}{\alpha}}+7e^{\frac{169\pi}{2\alpha}}+35e^{\frac{85\pi}{\alpha}}+3e^{\frac{171\pi}{2\alpha}}\\
&+10e^{\frac{86\pi}{\alpha}}+24e^{\frac{87\pi}{\alpha}}-7e^{\frac{175\pi}{2\alpha}}-5e^{\frac{88\pi}{\alpha}}-5e^{\frac{177\pi}{2\alpha}}-17e^{\frac{89\pi}{\alpha}}-8e^{\frac{179\pi}{2\alpha}}-24e^{\frac{90\pi}{\alpha}}\\
&+8e^{\frac{181\pi}{2\alpha}}-5e^{\frac{91\pi}{\alpha}}+25e^{\frac{183\pi}{2\alpha}}+23e^{\frac{92\pi}{\alpha}}+8e^{\frac{185\pi}{2\alpha}}+16e^{\frac{93\pi}{\alpha}}-30e^{\frac{187\pi}{2\alpha}}+8e^{\frac{94\pi}{\alpha}}\\
&-15e^{\frac{189\pi}{2\alpha}}+32e^{\frac{95\pi}{\alpha}}.
\end{aligned}
$$

$$
\begin{aligned}
B=&-2e^{\frac{5\pi}{2\alpha}}+12e^{\frac{7\pi}{2\alpha}}-e^{\frac{4\pi}{\alpha}}+4e^{\frac{9\pi}{2\alpha}}+e^{\frac{5\pi}{\alpha}}-3e^{\frac{11\pi}{2\alpha}}-e^{\frac{6\pi}{\alpha}}-11e^{\frac{13\pi}{2\alpha}}-10e^{\frac{15\pi}{2\alpha}}\\
&-2e^{\frac{8\pi}{\alpha}}+4e^{\frac{17\pi}{2\alpha}}+e^{\frac{9\pi}{\alpha}}+14e^{\frac{19\pi}{2\alpha}}+5e^{\frac{10\pi}{\alpha}}+3e^{\frac{21\pi}{2\alpha}}-e^{\frac{11\pi}{\alpha}}+15e^{\frac{23\pi}{2\alpha}}-9e^{\frac{12\pi}{\alpha}}\\
&-e^{\frac{25\pi}{2\alpha}}+e^{\frac{13\pi}{\alpha}}+13e^{\frac{27\pi}{2\alpha}}+12e^{\frac{29\pi}{2\alpha}}-2e^{\frac{15\pi}{\alpha}}-9e^{\frac{31\pi}{2\alpha}}-2e^{\frac{16\pi}{\alpha}}-6e^{\frac{33\pi}{2\alpha}}-6e^{\frac{35\pi}{2\alpha}}\\
&+9e^{\frac{18\pi}{\alpha}}+6e^{\frac{37\pi}{2\alpha}}+e^{\frac{19\pi}{\alpha}}+2e^{\frac{39\pi}{2\alpha}}+e^{\frac{20\pi}{\alpha}}+2e^{\frac{41\pi}{2\alpha}}+3e^{\frac{21\pi}{\alpha}}-12e^{\frac{43\pi}{2\alpha}}+10e^{\frac{22\pi}{\alpha}}\\
&-13e^{\frac{45\pi}{2\alpha}}-10e^{\frac{23\pi}{\alpha}}+10e^{\frac{47\pi}{2\alpha}}-10e^{\frac{24\pi}{\alpha}}-2e^{\frac{49\pi}{2\alpha}}+2e^{\frac{25\pi}{\alpha}}+2e^{\frac{51\pi}{2\alpha}}+14e^{\frac{53\pi}{2\alpha}}\\
&-4e^{\frac{27\pi}{\alpha}}+12e^{\frac{55\pi}{2\alpha}}-3e^{\frac{28\pi}{\alpha}}-10e^{\frac{57\pi}{2\alpha}}+3e^{\frac{29\pi}{\alpha}}+5e^{\frac{59\pi}{2\alpha}}+7e^{\frac{30\pi}{\alpha}}-7e^{\frac{61\pi}{2\alpha}}+2e^{\frac{31\pi}{\alpha}}\\
&-8e^{\frac{63\pi}{2\alpha}}-4e^{\frac{32\pi}{\alpha}}+8e^{\frac{65\pi}{2\alpha}}+5e^{\frac{33\pi}{\alpha}}+16e^{\frac{67\pi}{2\alpha}}+15e^{\frac{34\pi}{\alpha}}+9e^{\frac{69\pi}{2\alpha}}-3e^{\frac{35\pi}{\alpha}}+9e^{\frac{71\pi}{2\alpha}}\\
&-27e^{\frac{36\pi}{\alpha}}-3e^{\frac{73\pi}{2\alpha}}+3e^{\frac{37\pi}{\alpha}}+13e^{\frac{75\pi}{2\alpha}}+28e^{\frac{77\pi}{2\alpha}}-6e^{\frac{39\pi}{\alpha}}-5e^{\frac{79\pi}{2\alpha}}-4e^{\frac{40\pi}{\alpha}}\\
&-12e^{\frac{81\pi}{2\alpha}}-2e^{\frac{41\pi}{\alpha}}+2e^{\frac{83\pi}{2\alpha}}+17e^{\frac{42\pi}{\alpha}}-2e^{\frac{85\pi}{2\alpha}}+3e^{\frac{43\pi}{\alpha}}-16e^{\frac{87\pi}{2\alpha}}+3e^{\frac{44\pi}{\alpha}}\\
&-2e^{\frac{89\pi}{2\alpha}}+7e^{\frac{45\pi}{\alpha}}-10e^{\frac{91\pi}{2\alpha}}+20e^{\frac{46\pi}{\alpha}}-13e^{\frac{93\pi}{2\alpha}}-20e^{\frac{47\pi}{\alpha}}+12e^{\frac{95\pi}{2\alpha}}-20e^{\frac{48\pi}{\alpha}}\\
&-4e^{\frac{97\pi}{2\alpha}}+4e^{\frac{49\pi}{\alpha}}+28e^{\frac{99\pi}{2\alpha}}+32e^{\frac{101\pi}{2\alpha}}-8e^{\frac{51\pi}{\alpha}}+4e^{\frac{103\pi}{2\alpha}}-3e^{\frac{52\pi}{\alpha}}-12e^{\frac{105\pi}{2\alpha}}\\
&+3e^{\frac{53\pi}{\alpha}}+e^{\frac{107\pi}{2\alpha}}+9e^{\frac{54\pi}{\alpha}}-e^{\frac{109\pi}{2\alpha}}-28e^{\frac{111\pi}{2\alpha}}-6e^{\frac{56\pi}{\alpha}}-2e^{\frac{113\pi}{2\alpha}}+3e^{\frac{57\pi}{\alpha}}\\
&+18e^{\frac{115\pi}{2\alpha}}+19e^{\frac{58\pi}{\alpha}}-15e^{\frac{117\pi}{2\alpha}}-7e^{\frac{59\pi}{\alpha}}+25e^{\frac{119\pi}{2\alpha}}-67e^{\frac{60\pi}{\alpha}}-11e^{\frac{121\pi}{2\alpha}}+3e^{\frac{61\pi}{\alpha}}\\
&+25e^{\frac{123\pi}{2\alpha}}+4e^{\frac{62\pi}{\alpha}}+38e^{\frac{125\pi}{2\alpha}}-6e^{\frac{63\pi}{\alpha}}-11e^{\frac{127\pi}{2\alpha}}-2e^{\frac{64\pi}{\alpha}}-10e^{\frac{129\pi}{2\alpha}}-8e^{\frac{65\pi}{\alpha}}\\
&+4e^{\frac{131\pi}{2\alpha}}+31e^{\frac{66\pi}{\alpha}}-6e^{\frac{133\pi}{2\alpha}}+11e^{\frac{67\pi}{\alpha}}-28e^{\frac{135\pi}{2\alpha}}-e^{\frac{68\pi}{\alpha}}+12e^{\frac{137\pi}{2\alpha}}+21e^{\frac{69\pi}{\alpha}}\\
&-4e^{\frac{139\pi}{2\alpha}}+42e^{\frac{70\pi}{\alpha}}-37e^{\frac{141\pi}{2\alpha}}-42e^{\frac{71\pi}{\alpha}}+14e^{\frac{143\pi}{2\alpha}}-34e^{\frac{72\pi}{\alpha}}-10e^{\frac{145\pi}{2\alpha}}+10e^{\frac{73\pi}{\alpha}}
\end{aligned}
$$

$$+36e^{\frac{147\pi}{2\alpha}}-8e^{\frac{74\pi}{\alpha}}+52e^{\frac{149\pi}{2\alpha}}-20e^{\frac{75\pi}{\alpha}}+16e^{\frac{151\pi}{2\alpha}}-25e^{\frac{76\pi}{\alpha}}-14e^{\frac{153\pi}{2\alpha}}$$
$$+17e^{\frac{77\pi}{\alpha}}+9e^{\frac{155\pi}{2\alpha}}+65e^{\frac{78\pi}{\alpha}}-e^{\frac{157\pi}{2\alpha}}+14e^{\frac{79\pi}{\alpha}}+2e^{\frac{159\pi}{2\alpha}}-12e^{\frac{80\pi}{\alpha}}+6e^{\frac{161\pi}{2\alpha}}$$
$$+23e^{\frac{81\pi}{\alpha}}+12e^{\frac{163\pi}{2\alpha}}+41e^{\frac{82\pi}{\alpha}}-13e^{\frac{165\pi}{2\alpha}}-5e^{\frac{83\pi}{\alpha}}+19e^{\frac{167\pi}{2\alpha}}-41e^{\frac{84\pi}{\alpha}}-e^{\frac{169\pi}{2\alpha}}$$
$$+9e^{\frac{85\pi}{\alpha}}-7e^{\frac{171\pi}{2\alpha}}-4e^{\frac{86\pi}{\alpha}}+58e^{\frac{173\pi}{2\alpha}}-18e^{\frac{87\pi}{\alpha}}+5e^{\frac{175\pi}{2\alpha}}-8e^{\frac{88\pi}{\alpha}}-12e^{\frac{177\pi}{2\alpha}}$$
$$-6e^{\frac{89\pi}{\alpha}}-4e^{\frac{179\pi}{2\alpha}}+7e^{\frac{90\pi}{\alpha}}-2e^{\frac{181\pi}{2\alpha}}-7e^{\frac{91\pi}{\alpha}}-10e^{\frac{183\pi}{2\alpha}}+21e^{\frac{92\pi}{\alpha}}-12e^{\frac{185\pi}{2\alpha}}$$
$$-7e^{\frac{93\pi}{\alpha}}-34e^{\frac{187\pi}{2\alpha}}-16e^{\frac{94\pi}{\alpha}}+43e^{\frac{189\pi}{2\alpha}}$$

$$C=6+2e^{\frac{\pi}{\alpha}}-12e^{\frac{2\pi}{\alpha}}-8e^{\frac{3\pi}{\alpha}}+2e^{\frac{4\pi}{\alpha}}+12e^{\frac{5\pi}{\alpha}}+13e^{\frac{6\pi}{\alpha}}+5e^{\frac{7\pi}{\alpha}}-2e^{\frac{15\pi}{2\alpha}}-15e^{\frac{8\pi}{\alpha}}$$
$$-15e^{\frac{9\pi}{\alpha}}-7e^{\frac{10\pi}{\alpha}}-8e^{\frac{11\pi}{\alpha}}-30e^{\frac{12\pi}{\alpha}}-12e^{\frac{13\pi}{\alpha}}-2e^{\frac{27\pi}{2\alpha}}+18e^{\frac{14\pi}{\alpha}}+28e^{\frac{15\pi}{\alpha}}$$
$$+9e^{\frac{16\pi}{\alpha}}-e^{\frac{33\pi}{2\alpha}}-9e^{\frac{17\pi}{\alpha}}+e^{\frac{18\pi}{\alpha}}+2e^{\frac{19\pi}{\alpha}}+3e^{\frac{39\pi}{2\alpha}}-6e^{\frac{21\pi}{\alpha}}-8e^{\frac{22\pi}{\alpha}}+2e^{\frac{23\pi}{\alpha}}$$
$$+9e^{\frac{24\pi}{\alpha}}+8e^{\frac{25\pi}{\alpha}}+4e^{\frac{51\pi}{2\alpha}}-16e^{\frac{26\pi}{\alpha}}-18e^{\frac{27\pi}{\alpha}}-10e^{\frac{28\pi}{\alpha}}-2e^{\frac{57\pi}{2\alpha}}+12e^{\frac{29\pi}{\alpha}}$$
$$+21e^{\frac{30\pi}{\alpha}}+e^{\frac{31\pi}{\alpha}}-4e^{\frac{63\pi}{2\alpha}}-9e^{\frac{32\pi}{\alpha}}-9e^{\frac{33\pi}{\alpha}}-e^{\frac{34\pi}{\alpha}}-2e^{\frac{35\pi}{\alpha}}-32e^{\frac{36\pi}{\alpha}}-10e^{\frac{37\pi}{\alpha}}$$
$$-6e^{\frac{75\pi}{2\alpha}}+10e^{\frac{38\pi}{\alpha}}+12e^{\frac{39\pi}{\alpha}}+e^{\frac{40\pi}{\alpha}}-3e^{\frac{81\pi}{2\alpha}}-5e^{\frac{41\pi}{\alpha}}+11e^{\frac{42\pi}{\alpha}}-6e^{\frac{43\pi}{\alpha}}+5e^{\frac{87\pi}{2\alpha}}$$
$$-2e^{\frac{44\pi}{\alpha}}+11e^{\frac{45\pi}{\alpha}}-2e^{\frac{46\pi}{\alpha}}-8e^{\frac{47\pi}{\alpha}}-3e^{\frac{48\pi}{\alpha}}+2e^{\frac{49\pi}{\alpha}}+8e^{\frac{99\pi}{2\alpha}}-6e^{\frac{50\pi}{\alpha}}-2e^{\frac{51\pi}{\alpha}}$$
$$+2e^{\frac{52\pi}{\alpha}}-8e^{\frac{105\pi}{2\alpha}}+4e^{\frac{53\pi}{\alpha}}-2e^{\frac{107\pi}{2\alpha}}+e^{\frac{54\pi}{\alpha}}+2e^{\frac{109\pi}{2\alpha}}-3e^{\frac{55\pi}{\alpha}}-6e^{\frac{111\pi}{2\alpha}}+5e^{\frac{56\pi}{\alpha}}$$
$$+3e^{\frac{57\pi}{\alpha}}+4e^{\frac{115\pi}{2\alpha}}-3e^{\frac{58\pi}{\alpha}}+6e^{\frac{117\pi}{2\alpha}}-2e^{\frac{59\pi}{\alpha}}-2e^{\frac{119\pi}{2\alpha}}-44e^{\frac{60\pi}{\alpha}}+2e^{\frac{121\pi}{2\alpha}}$$
$$-10e^{\frac{61\pi}{\alpha}}-12e^{\frac{123\pi}{2\alpha}}+12e^{\frac{62\pi}{\alpha}}+22e^{\frac{63\pi}{\alpha}}+2e^{\frac{127\pi}{2\alpha}}-5e^{\frac{64\pi}{\alpha}}-7e^{\frac{129\pi}{2\alpha}}-5e^{\frac{65\pi}{\alpha}}$$
$$+4e^{\frac{131\pi}{2\alpha}}+29e^{\frac{66\pi}{\alpha}}+4e^{\frac{133\pi}{2\alpha}}-2e^{\frac{67\pi}{\alpha}}+5e^{\frac{135\pi}{2\alpha}}-20e^{\frac{68\pi}{\alpha}}-12e^{\frac{69\pi}{\alpha}}+8e^{\frac{139\pi}{2\alpha}}$$
$$-18e^{\frac{70\pi}{\alpha}}+18e^{\frac{141\pi}{2\alpha}}+6e^{\frac{71\pi}{\alpha}}+4e^{\frac{143\pi}{2\alpha}}+21e^{\frac{72\pi}{\alpha}}-4e^{\frac{145\pi}{2\alpha}}+12e^{\frac{73\pi}{\alpha}}+16e^{\frac{147\pi}{2\alpha}}$$
$$-14e^{\frac{74\pi}{\alpha}}-36e^{\frac{75\pi}{\alpha}}+4e^{\frac{151\pi}{2\alpha}}-22e^{\frac{76\pi}{\alpha}}+2e^{\frac{153\pi}{2\alpha}}+24e^{\frac{77\pi}{\alpha}}+2e^{\frac{155\pi}{2\alpha}}+5e^{\frac{78\pi}{\alpha}}$$
$$-2e^{\frac{157\pi}{2\alpha}}-11e^{\frac{79\pi}{\alpha}}-4e^{\frac{159\pi}{2\alpha}}+7e^{\frac{80\pi}{\alpha}}+21e^{\frac{81\pi}{\alpha}}-4e^{\frac{163\pi}{2\alpha}}+7e^{\frac{82\pi}{\alpha}}-2e^{\frac{165\pi}{2\alpha}}$$
$$+2e^{\frac{167\pi}{2\alpha}}-30e^{\frac{84\pi}{\alpha}}-2e^{\frac{169\pi}{2\alpha}}-12e^{\frac{85\pi}{\alpha}}-12e^{\frac{171\pi}{2\alpha}}+16e^{\frac{86\pi}{\alpha}}+14e^{\frac{87\pi}{\alpha}}-2e^{\frac{175\pi}{2\alpha}}$$
$$+7e^{\frac{88\pi}{\alpha}}-5e^{\frac{177\pi}{2\alpha}}-21e^{\frac{89\pi}{\alpha}}-8e^{\frac{179\pi}{2\alpha}}-17e^{\frac{90\pi}{\alpha}}-8e^{\frac{181\pi}{2\alpha}}-6e^{\frac{91\pi}{\alpha}}+11e^{\frac{183\pi}{2\alpha}}$$
$$+6e^{\frac{92\pi}{\alpha}}-8e^{\frac{185\pi}{2\alpha}}-37e^{\frac{93\pi}{\alpha}}-72e^{\frac{94\pi}{\alpha}}+10e^{\frac{189\pi}{2\alpha}}+64e^{\frac{95\pi}{\alpha}}+138e^{\frac{96\pi}{\alpha}}.$$

Appendix B
Implementation of Numerical Simulations

We have simulated numerically the dependence of the multiple-quantum cross-relaxation rate $\mu_{\mathrm{MQ}}^{\mathrm{WALTZ}}$ on the applied *rf* amplitude during a HDR WALTZ-32 pulse sequence under different exchange conditions.

The starting point for our simulations is the homogeneous master equation [1, 2] $\frac{d}{dt}\sigma(t) = -\hat{L}\sigma(t)$, where $\sigma(t)$ is the density operator of the two-spin system and the Liouvillian superoperator $\hat{L} = i\hat{H} + \hat{\Gamma}$ includes offsets, scalar couplings and the heteronuclear double-resonance *rf* irradiation, through the superoperator $\hat{H}$, as well as the effects of stochastic processes through the relaxation superoperator $\hat{\Gamma}$. The explicit matrix representation of the superoperators $\hat{H}$ and $\hat{\Gamma}$ can be found in [1]. The effects of chemical exchange are taken into account by modifying the master equation, much in the way as the classical Bloch equations were modified to give the McConnell equations [3].

We assume a simple first-order chemical exchange reaction between sites a and b, described by the kinetic matrix

$$K = \begin{bmatrix} -k_{\mathrm{f}} & k_{\mathrm{r}} \\ k_{\mathrm{f}} & -k_{\mathrm{r}} \end{bmatrix}. \tag{B.1}$$

where k_{f} and k_{r} are the forward and reverse exchange rates, respectively. The corresponding Liouville superoperator is obtained by the Kronecker product $\hat{\hat{K}} = K \otimes \hat{\mathscr{E}}_{16}$, where $\hat{\mathscr{E}}_{16}$ is the 16×16 unity superoperator. A new *product space* of dimension 32 is thus constructed by combining the 2×2 chemical configuration space and the 16×16 Liouville space in order to describe the jumps of the magnetization between the two exchanging sites [3, 4]. Henceforth double carets over operators will denote superoperators in the 32×32 product space. The Liouville superoperators $\hat{H}$ and $\hat{\Gamma}$ have to be adapted accordingly. Therefore, we define

N. Salvi, *Dynamic Studies Through Control of Relaxation in NMR Spectroscopy*, Springer Theses, DOI: 10.1007/978-3-319-06170-2,

$$\hat{\hat{H}}^k_\Omega = \begin{bmatrix} \omega^k_a & 0 \\ 0 & \omega^k_b \end{bmatrix} \otimes \hat{\mathscr{O}}^k_z \quad \text{with} \quad k = I, S, \tag{B.2a}$$

$$\hat{\hat{H}}_{\text{J}} = \hat{\mathscr{E}}_2 \otimes \hat{H}_{\text{J}}, \tag{B.2b}$$

$$\hat{\hat{H}}_{rf} = \hat{\mathscr{E}}_2 \otimes \hat{H}_{rf}, \tag{B.2c}$$

where $\hat{H}_{\text{J}}$ and $\hat{H}_{rf}$ are the Liouville superoperators associated with the scalar coupling between spins I and S and the rf fields, respectively, and $\hat{\mathscr{O}}^k_z$ $(k = I, S)$ are the dimensionless superoperators constructed from the (4×4) Zeeman spin operators I_z and S_z. How to obtain a matrix representation of Liouville superoperators using a basis set of Cartesian product operators is described elsewhere [1]. In a similar way one can modify the relaxation superoperator, assuming identical relaxation rates in states a and b,

$$\hat{\hat{\Gamma}} = \hat{\mathscr{E}}_2 \otimes \hat{\Gamma}. \tag{B.3}$$

The Liouville representation of the density operator is now a column vector of 32 elements and the master equation in the product space is

$$\frac{d}{dt}\sigma(t) = -\hat{\hat{L}}\sigma(t), \tag{B.4}$$

with $\hat{\hat{L}} = i\hat{\hat{H}} + \hat{\hat{\Gamma}} - \hat{\hat{K}}$ and $\hat{\hat{H}} = \hat{\hat{H}}^I_\Omega + \hat{\hat{H}}^S_\Omega + \hat{\hat{H}}_{\text{J}} + \hat{\hat{H}}_{rf}$. Over a time interval $[0, t]$ with a constant rf field the Liouvillian superoperator $\hat{\hat{L}}$ is time-independent in the DRF so that, given an initial state σ_0, the solution to Eq. B.4 at time t is

$$\sigma(t) = e^{-\hat{\hat{L}}t}\sigma_0. \tag{B.5}$$

We can set σ_0 equal to either $\mathbb{X} = 2I_x S_x$ or $\mathbb{Y} = 2I_y S_y$ and simulate the evolution of the system throughout the whole HDR WALTZ-32 sequence, including the effects of symmetrical reconversion [5, 6], which is employed to average out the difference of the effective auto-relaxation rates of the cross-relaxing operators [1]. Simulations are run for different relaxation delays, i.e., for n-fold repetitions of the HDR WALTZ-32 block. The effective cross-relaxation rate $\mu^{\text{WALTZ}}_{\text{MQ}}$ at a given rf amplitude ω_1 can be extracted by fitting the simulated data as explained elsewhere [1]. For the sake of simplicity, in this PhD thesis we show only results obtained for asymmetric populations $p_a = 0.98$ and $p_b = 0.02$. Moreover, we have assumed a scalar coupling constant $J_{IS} = -90\,\text{Hz}$, typical of ^{1}H-^{15}N systems in protein backbones, a rotational correlation time $\tau_c = 6\,\text{ns}$ and a static field $B_0 = 14.09\,\text{T}$, corresponding to a ^{1}H Larmor frequency of 600 MHz. In the simulations we assumed an uncertainty of 2 % on the exchange rate k_{ex} and on the chemical shift differences $\Delta\omega^I_{ab}$ and $\Delta\omega^S_{ab}$. These uncertainties were taken into account by a Montecarlo method.

References

1. Ulzega S, Verde M, Ferrage F, Bodenhausen G (2009) J Chem Phys 131:224503
2. Allard P, Helgstrand M, Hard T (1998) J Magn Reson 134:7
3. Helgstrand MM, Härd TT, Allard PP (2000) J Biomol NMR 18:49
4. Jeener J (1982) Adv Magn Reson 10:1
5. Pelupessy P, Espargallas GM, Bodenhausen G (2003) J Magn Reson 161:258
6. Pelupessy P, Ferrage F, Bodenhausen G (2007) J Chem Phys 126:134508

Appendix C
Implementation of Minimization Algorithms

C.1 Genetic Algorithm: Mathematica Version

```
GeneticAnalytic[Npop_, kexrange_, Δωrange_,
  data_, err_, ν1_] := Module[
  {chromTable (*each chromosome in a row*),
   cost = Table[0, {i, 1, Npop}],
   μinf = Table[0, {i, 1, Npop}],
   Nkeep = Round[Npop/2], (*50% selection rate*)
   Noffspr = Npop - Round[Npop/2],(*50% selection rate*)
   Ntourn = 3, (*for tournament selection*)
   NMut,
   MutRate = 0.25, (*25% mutation rate*)
   MaxIter = 50, j = 1,
   Prec = 5*10^(-2),
   sim, chi2, μ0, temp,
   mothers, fathers, tempM, tempF, tempC,
   MutRow, MutCol,
   kex, Δkex, Δω, ΔΔω, μinfFin, ΔμinfFin},
  (*definition of the initial population and cost*)
  chromTable =
   ParallelTable[{Random[Real, kexrange],
     Random[Real, Δωrange]}, {i, 1,
     Npop}]; (*this is Δω as defined in the CPC \
paper*)
  NMut = Round[MutRate*Noffspr];
  Do[
   sim = Table[
     HDRfunc[ν1[[i]], 1/chromTable[[j, 1]],
      chromTable[[j, 2]]], {i, 1, Length[ν1]}];
   chi2 =
    Sum[((data[[i]] - sim[[i]] - μ0)/err[[i]])^2, {i, 1,
      Length[ν1]}];
   temp = Minimize[chi2, μ0];
   μinf[[j]] = μ0 /. temp[[2]];
   cost[[j]] = temp[[1]];
   , {j, 1, Npop}];
  fathers = Table[{0, 0}, {i, 1, Noffspr}];
  mothers = Table[{0, 0}, {i, 1, Noffspr}];
  While[j < MaxIter + 1 &&
    StandardDeviation[Take[cost, Nkeep]]/Min[Take[cost, Nkeep]] > Prec,
   Print[j, "/", MaxIter, " , ",
    StandardDeviation[Take[cost, Nkeep]]/Min[Take[cost, Nkeep]]];
   (*Print[chromTable];*)
   (*natural selection*)
```

N. Salvi, *Dynamic Studies Through Control of Relaxation in NMR Spectroscopy*, Springer Theses, DOI: 10.1007/978-3-319-06170-2,

```
   chromTable =
    SortBy[chromTable, cost[[Flatten[Position[chromTable, #]]]] &];
   μinf =
    SortBy[μinf, cost[[Flatten[Position[μinf, #]]]] &];
   (*chromTable=Sort[chromTable,cost[[#1]]<cost[[#2]]&];*)

   cost = Sort[cost, #1 < #2 &];
   Do[chromTable[[Nkeep + i]] = {0, 0};
    μinf[[Nkeep + i]] = 0;
    cost[[Nkeep + i]] = 0
    , {i, 1, Noffspr}];
   (*tournament selection*)

   Do[tempF = Table[Random[Integer, {1, Nkeep}], {z, 1, Ntourn}];
    tempM = Table[Random[Integer, {1, Nkeep}], {z, 1, Ntourn}];
    tempC = Table[cost[[tempF[[z]]]], {z, 1, Ntourn}];
    fathers[[i]] = chromTable[[tempF[[Ordering[tempC, 1][[1]]]]]];
    tempC = Table[cost[[tempM[[z]]]], {z, 1, Ntourn}];
    mothers[[i]] = chromTable[[tempM[[Ordering[tempC, 1][[1]]]]]];
    , {i, 1, Noffspr}];
   (*mating*)

   Do[chromTable[[
       Nkeep + i]] = {(mothers[[i, 1]] + fathers[[i, 1]])/
        2, (mothers[[i, 2]] + fathers[[i, 2]])/2};
    , {i, 1, Noffspr}];
   (*mutations*)
   Do[MutRow = Random[Integer, {Nkeep + 1, Npop}];
    MutCol = Random[Integer, {1, 2}];
    If[MutCol == 1, chromTable[[MutRow, 1]] = Random[Real, kexrange],
     chromTable[[MutRow, 2]] =
      Random[Real, Δωrange]]
    , {i, 1, NMut}];
   (*cost of the new generation*)
   Do[
    sim =
     Table[HDRfunc[ν1[[i]], 1/chromTable[[Nkeep + z, 1]],
       chromTable[[Nkeep + z, 2]]], {i, 1, Length[ν1]}];
    chi2 =
     Sum[((data[[i]] - sim[[i]] - μ0)/err[[i]])^2, {i, 1,
       Length[ν1]}];
    temp = Minimize[chi2, μ0];
    μinf[[Nkeep + z]] = μ0 /. temp[[2]];
    cost[[Nkeep + z]] = temp[[1]];
    , {z, 1, Noffspr}];
   j++;
   ];

  chromTable =
   SortBy[chromTable,
    cost[[Flatten[Position[chromTable, #]]]] &]; μinf =
   SortBy[μinf, cost[[Flatten[Position[μinf, #]]]] &];
  cost = Sort[cost, #1 < #2 &];

  kex = chromTable[[1, 1]];(*Mean[chromTable[[All,
  1]]];*)
  Δkex =(*PDF[ChiSquareDistribution[
   Length[ν1]-4],cost[[1]]]*kex;*)
   StandardDeviation[chromTable[[1 ;; Nkeep, 1]]];
  Δω = chromTable[[1, 2]];(*Mean[chromTable[[All,
  2]]];*)
  ΔΔω =(*PDF[
   ChiSquareDistribution[Length[ν1]-4],cost[[
   1]]]*Δω;*)
   StandardDeviation[chromTable[[1 ;; Nkeep, 2]]];
  μinfFin = μinf[[
    1]];(*Mean[μinf];*)
  ΔμinfFin =(*PDF[
   ChiSquareDistribution[Length[ν1]-4],cost[[1]]]*μinfFin;*)
   StandardDeviation[Take[μinf, Nkeep]];
```

```
  {kex, Δkex, Δω, ΔΔω, μinfFin, ΔμinfFin}
  ]
```

```
(*Ulzega,S.;Salvi,N.;Segawa,T.F.;Ferrage,F.;Bodenhausen,G. ChemPhysChem
     2011,12,333*)

HDRfunc[ν1_, τex_, Δω2_] := Module[
  {corr = (23*(2 π*ν1)^3*τex^3)/(16*π*(1 -
          Exp[-96 π/(2*π*ν1*τex)]) (1 +
(2*π*ν1*τex)^2)), value},
  value = (Δω2*τex/(1 + (2*π*ν1* τex)^2))*(1 + corr);
  value]
```

C.2 Genetic Algorithm: Matlab Version

```
function [solution, STD] = GeneticAlgorithm(problem)

% =====================================================================
% DESCRIPTION
%
% usage: [solutions, values] = GeneticAlgorithm(problem)
%
% Solves the optimization problem "problem" using a GA.
% "problem" is a MATLAB data structure. Using a data strucure allows to
% pass problem specific information between the different function handles
% specified further below.
%
% ---------------------------------------------------------------------
% PARAMETERS
%
% problem.MAX_ITERATIONS      the maximum number of iterations
% problem.Accuracy            stop criterion
% problem.NumTournament       How many chromosomes in the tournament
%                             selection
% problem.NumChromosomes      Number of chromosomes
% problem.Ranges              Ranges for the variables
% problem.MutationRate        rate of the mutations
% problem.SelectionRate       selection rate (% that you keep)
% problem.OBJECTIVE_FUNCTION  a handle to a function that computes the
%                             objective function value for a given solution;
%                             this function must take the following
%                             parameters:
%                             1. the solution to be evaluated (a row vector)
%                             2. the "problem" data structure
%                             The return value of this function must be a
%                             scalar value.
% problem.SIMULATOR           a handle to a function that is used to
%                             simulate NMR data
%
% ---------------------------------------------------------------------
% RETURN VALUES
%
% solution                  the best solution
% value                     the corresponding value of the objective
%                           function
%
% =====================================================================

NumVariables=length(problem.Ranges);
Nkeep=round(problem.NumChromosomes*problem.SelectionRate);
NumOffSpring=problem.NumChromosomes-Nkeep;
NumMut=round(problem.MutationRate*NumOffSpring);

%initial population and cost

for i = 1:problem.NumChromosomes
    for j = 1:NumVariables
        ChromTable(i,j)=random('unif',problem.Ranges(j,1),problem.Ranges(j,2));
```

```
    end
    cost(i)=feval(problem.OBJECTIVE_FUNCTION,problem,ChromTable(i,:));
end

iter=1;
while iter < problem.MAX_ITERATIONS + 1 && std(cost(1:Nkeep))/min(cost(1:Nkeep)) > problem.Accuracy
    [cost, IX]=sort(cost);
    ChromTableStore=ChromTable;
    for i = 1:problem.NumChromosomes
        ChromTable(i,:)=ChromTableStore(IX(i),:);
    end

    %tournament selection
    for i = 1:NumOffSpring
        tempF=randi(Nkeep,[1,problem.NumTournament]);
        tempM=randi(Nkeep,[1,problem.NumTournament]);
        for j=1:problem.NumTournament
            tempC(j)=cost(tempF(j));
        end
        [~, IX]=sort(tempC);
        fathers(i,:)=ChromTable(tempF((IX(1))),:);
        for j=1:problem.NumTournament
            tempC(j)=cost(tempM(j));
        end
        [~, IX]=sort(tempC);
        mothers(i,:)=ChromTable(tempM((IX(1))),:);
    end

    %mating
    for i = 1:NumOffSpring
        ChromTable(Nkeep+i,:)=(fathers(i,:)+mothers(i,:))/2;
    end

    %mutations
    for i = 1:NumMut
        MutRow=randi([Nkeep+1,problem.NumChromosomes],[1,1]);
        MutCol=randi(NumVariables,[1,1]);
        ChromTable(MutRow,MutCol)=random('unif',problem.Ranges(MutCol,1),problem.Ranges(MutCol,2));

    end

    %cost of the new generation
    for i = 1:NumOffSpring
        cost(Nkeep+i)=feval(problem.OBJECTIVE_FUNCTION,problem,ChromTable(Nkeep+i,:));
    end

    iter=iter+1;
end

%final sorting
[cost, IX]=sort(cost);
ChromTableStore=ChromTable;
    for i = 1:problem.NumChromosomes
        ChromTable(i,:)=ChromTableStore(IX(i),:);
    end

%output
solution=ChromTable(1,:);
STD=std(ChromTable(1:Nkeep,:),0,2);

end
```

C.3 Simulated Annealing: Matlab Version

```
function [solutions, values] = SimulatedAnnealing(problem)

% ===========================================================================
% DESCRIPTION
%
% usage: [solutions, values] = SimulatedAnnealing(problem)
%
% Solves the optimization problem "problem" using simulated annealing.
% "problem" is a MATLAB data structure. Using a data strucure allows to
% pass problem specific information between the different function handles
% specified further below.
%
% ---------------------------------------------------------------------------
% PARAMETERS
%
% problem.MAX_ITERATIONS        the maximum number of iterations
% problem.C                     C * log(iteration + 1) scales the difference
%                               between a new and the current solution
%                               (also called "C" in the lecture)
% problem.INITIAL_SOLUTION      a row vector that represents an initial solution
%                               of the problem
% problem.RANDOMIZE             a handle to a function that generates a proposal
%                               solution from a given solution; this function
%                               must take the following two parameters:
%                               1. the current solution (a row vector)
%                               2. the entire "problem" data structure
%                               The return value of this function must again be
%                               a row vector of appropriate dimension.
% problem.OBJECTIVE_FUNCTION    a handle to a function that computes the
%                               objective function value for a given solution;
%                               this function must take the following
%                               parameters:
%                               1. the solution to be evaluated (a row vector)
%                               2. the "problem" data structure
%                               The return value of this function must be a
%                               scalar value.
% problem.SIMULATOR             a handle to a function that is used to
%                               simulate NMR data
%
% ---------------------------------------------------------------------------
% RETURN VALUES
%
% solutions                     a matrix where the i-th row contains the
%                               solution computed in the i-th iteration
% values                        a column vector where the i-th row contains
%                               the objective function value computed in the
%                               i-th iteration
%
% ===========================================================================

x = problem.INITIAL_SOLUTION;
solutions(1,:) = x;
fx = feval(problem.OBJECTIVE_FUNCTION,problem,x);
values(1) = fx;
y = feval(problem.RANDOMIZE,x);
fy = feval(problem.OBJECTIVE_FUNCTION,problem,y);

lambda = problem.C*log(1+0.5);
%fprintf('SIMULATED ANNEALING ITERATION %1 \n \n \n \n',0)

for k = 2:problem.MAX_ITERATIONS

%       [fx, AvgMaxQueue_x, AvgTotal_x] = feval(problem.OBJECTIVE_FUNCTION,x,
        oldScenario);
        numer = exp(-lambda*fy);
        denom  = exp(-lambda*fx);
        alpha = min(1,numer/denom);

        r = rand();
```

```



    if r < alpha

        solutions(k,:) = y;
        values(k) = fy;

        fx = fy;
        x = y;

        y = feval(problem.RANDOMIZE,x);
        fy = feval(problem.OBJECTIVE_FUNCTION,problem,y);
        lambda = problem.C*log(1+k);


    else
        solutions(k,:) = x;
        values(k) = fx;


        y = feval(problem.RANDOMIZE,x);
        fy = feval(problem.OBJECTIVE_FUNCTION,problem,y);
        lambda = problem.C*log(1+k);


    end

    %fprintf('SIMULATED ANNEALING ITERATION %i \n \n \n \n',k)

end

end
```

Publications

1. Salvi, N., Frey, J., Carnevale, D., Grätzel, M., Bodenhausen, G. "Solid-state NMR of ruthenium sensitizers adsorbed on TiO_2". *Dalton Trans.* (2014), 43, 6389.
2. Salvi, N. "Dynamic Studies through Control of Relaxation in NMR Spectroscopy". Springer (2014).
3. Bornet, A., Milani, J., Wang, S., Mammoli, D., Buratto, R., Salvi, N., Segawa, T.F., Vizthum, V., Miéville, P., Chinthalapalli, S., Perez-Linde, A.J., Carnevale, D., Jannin, S., Caporini, M.A., Ulzega, S., Rey, M., Bodenhausen, G. "Dynamic Nuclear Polarization and Other Magnetic Ideas at EPFL". *Chimia* (2012), 66, 734.
4. Salvi, N., Buratto, R., Bornet, A., Ulzega, S., Rentero Rebollo, I., Angelini, A., Heinis, C., Bodenhausen, G. "Boosting Sensitivity of Ligand-Protein Screening by NMR of Long-Lived States". *J. Am. Chem. Soc.* (2012), 134, 11076.
5. Salvi, N., Ulzega, S., Ferrage, F., Bodenhausen, G. "Time Scales of Slow Motions in Ubiquitin Explored by Heteronuclear Double Resonance". *J. Am. Chem. Soc.* (2012), 134, 2481.
6. Segawa, T.F., Bornet, A., Salvi, N., Mièville, P., Vizthum, V., Carnevale, D., Jannin, S., Caporini, M.A., Ulzega, S., Vasos, P.R., Rey, M., Bodenhausen, G. "Extending Timescales and Narrowing Linewidths in NMR". *Chimia* (2011), 65, 652.
7. Ulzega, S., Salvi, N., Segawa, T.F., Ferrage, F., Bodenhausen, G. "Control of Cross Relaxation of Multiple-Quantum Coherences Induced by Fast Chemical Exchange under Heteronuclear Double-Resonance Irradiation". *ChemPhysChem* (2011), 12, 333.
8. Salvi, N., Belpassi, L., Zuccaccia D., Tarantelli, F., Macchioni, A. "Ion pairing in NHC gold(I) olefin complexes: a combined experimental/theoretical study". *J. Organomet. Chem.* (2010), 695, 2679.
9. Salvi, N., Belpassi, L., Tarantelli, F. "On the Dewar-Chatt-Duncanson model for catalytic gold(I) complexes". *Chem. Eur. J.* (2010), 16, 7231.

N. Salvi, *Dynamic Studies Through Control of Relaxation in NMR Spectroscopy*,
Springer Theses, DOI: 10.1007/978-3-319-06170-2,

In Review or in Preparation

1. Salvi, N. "Tools for the design and the interpretation of NMR relaxation dispersion experiments". *Prog. Nucl. Mag. Reson. Spectrosc., Invited article.*
2. Augustyniak, R., Salvi, N., Khan, S.N., Pelupessy, P., Bodenhausen, G., Lequin, O., Ferrage, F. "Multiple-Field Nitrogen-15 Relaxation Reveals Structure and Dynamics of the Partially Disordered Protein Engrailed". *In preparation.*
3. Buratto, R., Bornet, A., Salvi, N., Laguerre, A., Passemard, S., GerberLemaire, S., Bodenhausen, G. "Spin-Pair Labels to Extend Protein-Ligand Screening using Long-Lived Coherences and Long-Lived States". *Submitted.*

Patents

1. F04296: Method for the NMR based determination of the affinity of drugs for a target protein. A. Bornet, R. Buratto, N. Salvi, G. Bodenhausen. Date of filing European Patent: 30.11.2011

Zeitfracht Medien GmbH
Ferdinand-Jühlke-Straße 7
99095 Erfurt, Deutschland
produktsicherheit@kolibri360.de